Learn, Practice, Succeed

Eureka Math®
Grade 6
Module 3

Published by Great Minds®.

Copyright © 2019 Great Minds®.

Printed in the U.S.A.

This book may be purchased from the publisher at eureka-math.org.

4 5 6 7 8 9 10 LSC 26 25 24 23 22 21

ISBN 978-1-64054-966-1

G6-M3-LPS-05.2019

Students, families, and educators:

Thank you for being part of the *Eureka Math*® community, where we celebrate the joy, wonder, and thrill of mathematics.

In *Eureka Math* classrooms, learning is activated through rich experiences and dialogue. That new knowledge is best retained when it is reinforced with intentional practice. The *Learn, Practice, Succeed* book puts in students' hands the problem sets and fluency exercises they need to express and consolidate their classroom learning and master grade-level mathematics. Once students learn and practice, they know they can succeed.

What is in the Learn, Practice, Succeed *book?*

Fluency Practice: Our printed fluency activities utilize the format we call a Sprint. Instead of rote recall, Sprints use patterns across a sequence of problems to engage students in reasoning and to reinforce number sense while building speed and accuracy. Sprints are inherently differentiated, with problems building from simple to complex. The tempo of the Sprint provides a low-stakes adrenaline boost that increases memory and automaticity.

Classwork: A carefully sequenced set of examples, exercises, and reflection questions support students' in-class experiences and dialogue. Having classwork preprinted makes efficient use of class time and provides a written record that students can refer to later.

Exit Tickets: Students show teachers what they know through their work on the daily Exit Ticket. This check for understanding provides teachers with valuable real-time evidence of the efficacy of that day's instruction, giving critical insight into where to focus next.

Homework Helpers and Problem Sets: The daily Problem Set gives students additional and varied practice and can be used as differentiated practice or homework. A set of worked examples, Homework Helpers, support students' work on the Problem Set by illustrating the modeling and reasoning the curriculum uses to build understanding of the concepts the lesson addresses.

Homework Helpers and Problem Sets from prior grades or modules can be leveraged to build foundational skills. When coupled with *Affirm*®, *Eureka Math*'s digital assessment system, these Problem Sets enable educators to give targeted practice and to assess student progress. Alignment with the mathematical models and language used across *Eureka Math* ensures that students notice the connections and relevance to their daily instruction, whether they are working on foundational skills or getting extra practice on the current topic.

Where can I learn more about Eureka Math *resources?*

The Great Minds® team is committed to supporting students, families, and educators with an ever-growing library of resources, available at eureka-math.org. The website also offers inspiring stories of success in the *Eureka Math* community. Share your insights and accomplishments with fellow users by becoming a *Eureka Math* Champion.

Best wishes for a year filled with "aha" moments!

Jill Diniz

Jill Diniz
Chief Academic Officer, Mathematics
Great Minds

Contents

Module 3: Rational Numbers

Exploratory Challenge: Constructing the Number Line

Exercises

Complete the diagrams. Count by ones to label the number lines.

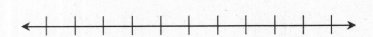

1. Plot your point on both number lines.

2. Show and explain how to find the opposite of your number on both number lines.

3. Mark the opposite on both number lines.

4. Choose a group representative to place the opposite number on the class number lines.

5. Which group had the opposite of the number on your index card?

EUREKA
MATH®

Name _____   Date _____

1. If zero lies between a and d, give one set of possible values for a, b, c, and d.

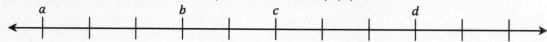

2. Below is a list of numbers in order from least to greatest. Use what you know about the number line to complete the list of numbers by filling in the blanks with the missing integers.

    −6, −5, _____, −3, −2, −1, _____, 1, 2, _____, 4, _____, 6

3. Complete the number line scale. Explain and show how to find 2 and the opposite of 2 on a number line.

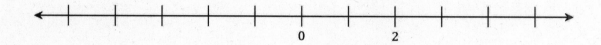

1. Draw a number line, and create a scale for the number line in order to plot the points −1, 3, and 5.

   a. Graph each point and its opposite on the number line.

   b. Explain how you found the opposite of each point.

> I know that opposite numbers are the same distance from zero, except in opposite directions.

To graph each point, start at zero, and move right or left based on the sign and number (to the right for a positive number and to the left for a negative number). To graph the opposites, start at zero, but this time move in the opposite direction the same number of times.

2. Kip uses a vertical number line to graph the points −3, −1, 2, and 5. He notices −3 is closer to zero than −1. He is not sure about his diagram. Use what you know about a vertical number line to determine if Kip made a mistake or not. Support your explanation with a vertical number line diagram.

   Kip made a mistake because −3 is less than −1, so it should be farther down the number line. Starting at zero, negative numbers decrease as we look farther below zero. So, −1 lies before −3 since −1 is 1 unit below zero, and −3 is three units below zero.

> I know that values increase as I look up on a vertical number line and decrease as I look down. Numbers above zero are positive, and numbers below zero are negative.

3. Create a scale in order to graph the numbers −10 through 10 on a number line. What does each tick mark represent?

    Each tick mark represents one unit.

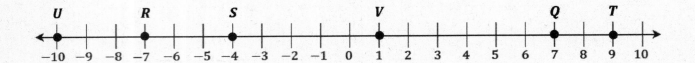

4. Choose an integer between −4 and −9. Label it R on the number line created in Problem 3, and complete the following tasks.

    Answers will vary. Answers a-e reflect the student choice of −7. −7 is between −4 and −9.

    a. What is the opposite of R? Label it Q.

        The opposite of −7 is 7.

    b. State a positive integer greater than Q. Label it T.

        A positive integer greater than 7 is 9 because 9 is farther to the right on the number line.

    c. State a negative integer greater than R. Label it S.

        A negative integer greater than −7 is −4 because −4 is farther to the right on the number line.

    d. State a negative integer less than R. Label it U.

        A negative integer less than −7 is −10 because −10 is farther to the left on the number line.

    e. State an integer between R and Q. Label it V.

        An integer between −7 and 7 is 1.

5. Will the opposite of a positive number always, sometimes, or never be a positive number? Explain your reasoning.

    The opposite of a positive number will never be a positive number. For two nonzero numbers to be opposite, zero has to be between both numbers, and the distance from zero to one number has to equal the distance between zero and the other number.

Lesson 1:   Positive and Negative Numbers on the Number Line—
                            Opposite Direction and Value

                    © 2019 Great Minds®. eureka-math.org

EUREKA
MATH

6. Will the opposite of zero always, sometimes, or never be zero? Explain your reasoning.

   The opposite of zero will always be zero because zero is its own opposite.

7. Will the opposite of a number always, sometimes, or never be greater than the number itself? Explain your reasoning. Provide an example to support your reasoning.

   The opposite of a number will sometimes be greater than the number itself because it depends on the given number. The opposite of a negative number is a positive number, so the opposite will be greater. But, the opposite of a positive number is a negative number, which is not greater. Also, if the number given is zero, then the opposite is zero, which is never greater than itself.

1. Draw a number line, and create a scale for the number line in order to plot the points −2, 4, and 6.

   a. Graph each point and its opposite on the number line.

   b. Explain how you found the opposite of each point.

2. Carlos uses a vertical number line to graph the points −4, −2, 3, and 4. He notices that −4 is closer to zero than −2. He is not sure about his diagram. Use what you know about a vertical number line to determine if Carlos made a mistake or not. Support your explanation with a number line diagram.

3. Create a scale in order to graph the numbers −12 through 12 on a number line. What does each tick mark represent?

4. Choose an integer between −5 and −10. Label it R on the number line created in Problem 3, and complete the following tasks.

   a. What is the opposite of R? Label it Q.

   b. State a positive integer greater than Q. Label it T.

   c. State a negative integer greater than R. Label it S.

   d. State a negative integer less than R. Label it U.

   e. State an integer between R and Q. Label it V.

5. Will the opposite of a positive number always, sometimes, or never be a positive number? Explain your reasoning.

6. Will the opposite of zero always, sometimes, or never be zero? Explain your reasoning.

7. Will the opposite of a number always, sometimes, or never be greater than the number itself? Explain your reasoning. Provide an example to support your reasoning.

Example 1: Take It to the Bank

Read Example 1 silently. In the first column, write down any words and definitions you know. In the second column, write down any words you do not know.

For Tim's 13th birthday, he received $150 in cash from his mom. His dad took him to the bank to open a savings account. Tim gave the cash to the banker to deposit into the account. The banker credited Tim's new account $150 and gave Tim a receipt. One week later, Tim deposited another $25 that he had earned as allowance. The next month, Tim's dad gave him permission to withdraw $35 to buy a new video game. Tim's dad explained that the bank would charge a $5 fee for each withdrawal from the savings account and that each withdrawal and charge results in a debit to the account.

Words I <u>Already Know</u>:	Words I <u>Want to Know</u>:	Words I <u>Learned</u>:

In the third column, write down any new words and definitions that you learn during the discussion.

© 2019 Great Minds®. eureka-math.org

Exercises 1–2

1. Read Example 1 again. With your partner, number the events in the story problem. Write the number above each sentence to show the order of the events.

For Tim's 13th birthday, he received $150 in cash from his mom. His dad took him to the bank to open a savings account.

Tim gave the cash to the banker to deposit into the account. The banker credited Tim's new account $150 and gave Tim

a receipt. One week later, Tim deposited another $25 that he had earned as allowance. The next month, Tim's dad gave

him permission to withdraw $35 to buy a new video game. Tim's dad explained that the bank would charge a $5 fee for

each withdrawal from the savings account and that each withdrawal and charge results in a debit to the account.

2. Write each individual description below as an integer. Model the integer on the number line using an appropriate scale.

EVENT	INTEGER	NUMBER LINE MODEL
Open a bank account with $0.		
Make a $150 deposit.		
Credit an account for $150.		
Make a deposit of $25.		
A bank makes a charge of $5.		
Tim withdraws $35.		

EUREKA MATH®

Example 2: How Hot, How Cold?

Temperature is commonly measured using one of two scales, Celsius or Fahrenheit. In the United States, the Fahrenheit system continues to be the accepted standard for nonscientific use. All other countries have adopted Celsius as the primary scale in use. The thermometer shows how both scales are related.

a. The boiling point of water is 100°C. Where is 100 degrees Celsius located on the thermometer to the right?

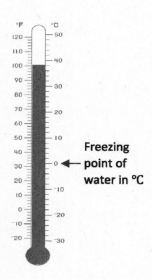

Freezing point of water in °C

b. On a vertical number line, describe the position of the integer that represents 100°C.

c. Write each temperature as an integer.
   i. The temperature shown on the thermometer in degrees Fahrenheit:

   ii. The temperature shown on the thermometer in degrees Celsius:

   iii. The freezing point of water in degrees Celsius:

d. If someone tells you your body temperature is 98.6°, what scale is being used? How do you know?

e. Does the temperature 0 degrees mean the same thing on both scales?

Exercises 3–5

3. Write each word under the appropriate column, "Positive Number" or "Negative Number."

    Gain   Loss   Deposit   Credit   Debit   Charge   Below Zero   Withdraw   Owe   Receive

Positive Number	Negative Number

4. Write an integer to represent each of the following situations:

    a. A company loses $345,000 in 2011.   _____

    b. You earned $25 for dog sitting.   _____

    c. Jacob owes his dad $5.   _____

    d. The temperature at the sun's surface is about 5,500°C.   _____

    e. The temperature outside is 4 degrees below zero.   _____

    f. A football player lost 10 yards when he was tackled.   _____

5. Describe a situation that can be modeled by the integer -15. Explain what zero represents in the situation.

EUREKA MATH®

Name _____   Date _____

1. Write a story problem that includes both integers −8 and 12.

2. What does zero represent in your story problem?

3. Choose an appropriate scale to graph both integers on the vertical number line. Label the scale.

4. Graph both points on the vertical number line.

1. Express each situation as an integer in the space provided.

    a. A gain of 45 points in a game

    45

    > I know words that imply a positive magnitude include "gain" and "deposit." Words that imply a negative magnitude include "fee charged," "below zero," and "loss."

    b. A fee charged of $3

    −3

    c. A temperature of 20 degrees Celsius below zero

    −20

    d. A 35-yard loss in a football game

    −35

    e. A $15,000 deposit

    15,000

2. Each sentence is stated *incorrectly*. Rewrite the sentence to correctly describe each situation.

    a. The temperature is −20 degrees Fahrenheit below zero.

    The temperature is* 20 *degrees Fahrenheit below zero. Or, the temperature is* −20 *degrees Fahrenheit.

b.   The temperature is −32 degrees Celsius below zero.

The temperature is 32 degrees Celsius below zero. Or, the temperature is −32 degrees Celsius.

I know that magnitude can be determined by the use of language in a problem. "Below zero" means that the number being referenced will be negative, so I do not need to use a negative sign. Or, if I choose to use a negative sign, I do not need the term "below zero" because the number is already negative.

For Problems 3–5, use the thermometer to the right.

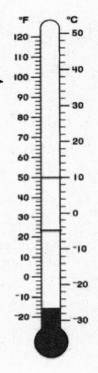

3.   Mark the integer on the thermometer that corresponds to the temperature given.
   a.   50°F
   b.   −5°C

The Fahrenheit scale is on the left of the thermometer, and the Celsius scale is on the right. I need to mark the integers on the correct scale.

4.   The melting point of steel is 1,510°C. Can this thermometer be used to record the temperature of the melting point of steel? Explain.

The melting point of steel cannot be represented on this thermometer. The highest this thermometer gauges is 50°C. 1, 510°C is a much larger value.

5.   Natalie shaded the thermometer to represent a temperature of 15 degrees below zero Celsius, as shown in the diagram. Is she correct? Why or why not? If necessary, describe how you would fix Natalie's shading.

Natalie is incorrect. She did shade in −15° but on the wrong scale. The shading represents −15°F, instead of −15°C. To fix Natalie's mistake, the shading must be between −10 and −20 on the Celsius scale.

EUREKA
MATH

1. Express each situation as an integer in the space provided.

   a. A gain of 56 points in a game

   _____

   b. A fee charged of $2

   _____

   c. A temperature of 32 degrees below zero

   _____

   d. A 56-yard loss in a football game

   _____

   e. The freezing point of water in degrees Celsius

   _____

   f. A $12,500 deposit

   _____

For Problems 2–5, use the thermometer to the right.

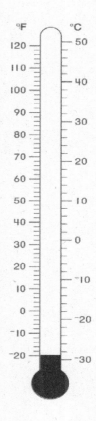

2. Each sentence is stated *incorrectly*. Rewrite the sentence to correctly describe each situation.

   a. The temperature is −10 degrees Fahrenheit below zero.

   b. The temperature is −22 degrees Celsius below zero.

3. Mark the integer on the thermometer that corresponds to the temperature given.

   a. 70°F

   b. 12°C

   c. 110°F

   d. −4°C

4. The boiling point of water is 212°F. Can this thermometer be used to record the temperature of a boiling pot of water? Explain.

5. Kaylon shaded the thermometer to represent a temperature of 20 degrees below zero Celsius as shown in the diagram. Is she correct? Why or why not? If necessary, describe how you would fix Kaylon's shading.

Example 1: A Look at Sea Level

The picture below shows three different people participating in activities at three different elevations. With a partner, discuss what you see. What do you think the word *elevation* means in this situation?

Exercises

Refer back to Example 1. Use the following information to answer the questions.

- The scuba diver is 30 feet below sea level.
- The sailor is at sea level.
- The hiker is 2 miles (10,560 feet) above sea level.

EUREKA MATH

1. Write an integer to represent each situation.

2. Use an appropriate scale to graph each of the following situations on the number line to the right. Also, write an integer to represent both situations.

   a. A hiker is 15 feet above sea level.

   b. A diver is 20 feet below sea level.

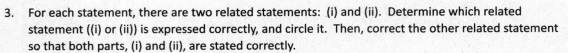

3. For each statement, there are two related statements: (i) and (ii). Determine which related statement ((i) or (ii)) is expressed correctly, and circle it. Then, correct the other related statement so that both parts, (i) and (ii), are stated correctly.

   a. A submarine is submerged 800 feet below sea level.

      i. The depth of the submarine is −800 feet below sea level.

      ii. 800 feet below sea level can be represented by the integer −800.

   b. The elevation of a coral reef with respect to sea level is given as −150 feet.

      i. The coral reef is 150 feet below sea level.

      ii. The depth of the coral reef is −150 feet below sea level.

Lesson 3:   Real-World Positive and Negative Numbers and Zero

EUREKA MATH

Name _____   Date _____

Exploratory Challenge Station Record Sheet

Poster # _____

Integers: _____

Number Line Scale: _____

Poster # _____

Integers: _____

Number Line Scale: _____

Poster # _____

Integers: _____

Number Line Scale: _____

Poster # _____

Integers: _____

Number Line Scale: _____

Poster # _____

Integers: _____

Number Line Scale: _____

#1   #2   #3   #4   #5

Name _____   Date _____

1. Write a story problem using sea level that includes both integers −110 and 120.

2. What does zero represent in your story problem?

3. Choose an appropriate scale to graph both integers on the vertical number line.

4. Graph and label both points on the vertical number line.

1. Write an integer to match the following descriptions.

   a. A debit of $50   **−50**

   b. A deposit of $125   **125**

   c. 5, 600 feet above sea level   **5, 600**

   d. A temperature increase of 50°F   **50**

   e. A withdrawal of $125   **−125**

   f. 5, 600 feet below sea level   **−5, 600**

> I know words that describe positive integers include "deposit," "above sea level," and "increase." Words that describe negative integers include "debit," "withdrawal," and "below sea level."

For Problems 2 and 3, read each statement about a real-world situation and the two related statements in parts (a) and (b) carefully. Circle the correct way to describe each real-world situation; *possible answers include either (a), (b), or both (a) and (b).*

2. A shark is 500 feet below the surface of the ocean.

   a. (The depth of the shark is 500 feet from the ocean's surface.)

   b. The whale is −500 feet below the surface of the ocean.

> To represent a negative integer, I know I can use a negative sign or vocabulary that determines magnitude, but not both.

3. Carl's body temperature decreased by 3°F.

   a. (Carl's body temperature dropped 3°F.)

   b. (The integer −3 represents the change in Carl's body temperature in degrees Fahrenheit.)

> The word "dropped" tells me the integer is negative. A "decrease" also tells me the integer is negative. I know that −3 represents a negative integer and the change in the temperature, so both of these examples are correct.

4. A credit of $45 and a debit of $50 are applied to your bank account.

    a. What is the appropriate scale to graph a credit of $45 and a debit of $50? Explain your reasoning.

       Because both numbers are divisible by 5, an interval of 5 is an appropriate scale on a number line.

    b. What integer represents "a credit of $45" if zero represents the original balance? Explain.

       45; a credit is greater than zero, and numbers greater than zero are positive numbers.

    c. What integer describes "a debit of $50" if zero represents the original balance? Explain.

       −50; a debit is less than zero, and numbers less than zero are negative numbers.

    d. Based on your scale, describe the location of both integers on the number line.

       If the scale is created with multiples of 5, then 45 would be 9 units to the right (or above) zero, and −50 would be 10 units to the left (or below) zero.

    e. What does zero represent in this situation?

       Zero represents no change being made to the account balance. No amount is either added to or subtracted from the account.

EUREKA MATH®

1. Write an integer to match the following descriptions.

    a.   A debit of $40   _____

    b.   A deposit of $225   _____

    c.   14,000 feet above sea level   _____

    d.   A temperature increase of 40°F   _____

    e.   A withdrawal of $225   _____

    f.   14,000 feet below sea level   _____

For Problems 2–4, read each statement about a real-world situation and the two related statements in parts (a) and (b) carefully. Circle the correct way to describe each real-world situation; *possible answers include either (a), (b), or both (a) and (b).*

2. A whale is 600 feet below the surface of the ocean.

    a.   The depth of the whale is 600 feet from the ocean's surface.

    b.   The whale is −600 feet below the surface of the ocean.

3. The elevation of the bottom of an iceberg with respect to sea level is given as −125 feet.

    a.   The iceberg is 125 feet above sea level.

    b.   The iceberg is 125 feet below sea level.

4. Alex's body temperature decreased by 2°F.

    a.   Alex's body temperature dropped 2°F.

    b.   The integer −2 represents the change in Alex's body temperature in degrees Fahrenheit.

5. A credit of $35 and a debit of $40 are applied to your bank account.

    a.   What is an appropriate scale to graph a credit of $35 and a debit of $40? Explain your reasoning.

    b.   What integer represents "a credit of $35" if zero represents the original balance? Explain.

    c.   What integer describes "a debit of $40" if zero represents the original balance? Explain.

    d.   Based on your scale, describe the location of both integers on the number line.

    e.   What does zero represent in this situation?

Exercise 1: Walk the Number Line

1. Each nonzero integer has an opposite, denoted $-a$; $-a$ and a are opposites if they are on opposite sides of zero and the same distance from zero on the number line.

Example 1: Every Number Has an Opposite

Locate the number 8 and its opposite on the number line. Explain how they are related to zero.

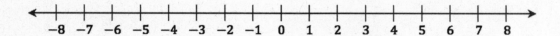

Exercises 2–3

2. Locate and label the opposites of the numbers on the number line.

    a. 9

    b. −2

    c. 4

    d. −7

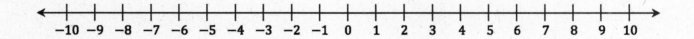

3. Write the integer that represents the opposite of each situation. Explain what zero means in each situation.

   a. 100 feet above sea level

   b. 32°C below zero

   c. A withdrawal of $25

Maria decides to take a walk along Central Avenue to purchase a book at the bookstore. On her way, she passes the Furry Friends Pet Shop and goes in to look for a new leash for her dog. Furry Friends Pet Shop is seven blocks west of the bookstore. She leaves Furry Friends Pet Shop and walks toward the bookstore to look at some books. After she leaves the bookstore, she heads east for seven blocks and stops at Ray's Pet Shop to see if she can find a new leash at a better price. Which location, if any, is the farthest from Maria while she is at the bookstore?

Determine an appropriate scale, and model the situation on the number line below.

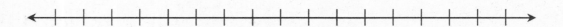

Explain your answer. What does zero represent in the situation?

Exercises 4–6

Read each situation carefully, and answer the questions.

4. On a number line, locate and label a credit of $15 and a debit for the same amount from a bank account. What does zero represent in this situation?

5. On a number line, locate and label 20°C below zero and 20°C above zero. What does zero represent in this situation?

6. A proton represents a positive charge. Write an integer to represent 5 protons. An electron represents a negative charge. Write an integer to represent 3 electrons.

Name _____   Date _____

In a recent survey, a magazine reported that the preferred room temperature in the summer is 68°F. A wall thermostat, like the ones shown below, tells a room's temperature in degrees Fahrenheit.

Sarah's Upstairs Bedroom   Downstairs Bedroom

a. Which bedroom is warmer than the recommended room temperature?

b. Which bedroom is cooler than the recommended room temperature?

c. Sarah notices that her room's temperature is 4°F above the recommended temperature, and the downstairs bedroom's temperature is 4°F below the recommended temperature. She graphs 72 and 64 on a vertical number line and determines they are opposites. Is Sarah correct? Explain.

d. After determining the relationship between the temperatures, Sarah now decides to represent 72°F as 4 and 64°F as −4 and graphs them on a vertical number line. Graph 4 and −4 on the vertical number line on the right. Explain what zero represents in this situation.

0

1. Find the opposite of each number, and describe its location on the number line.

    a. −4

        The opposite of −4 is 4, which is 4 units to the right of (or above) zero if the scale is one.

    b. 8

        The opposite of 8 is −8, which is 8 units to the left of (or below) zero if the scale is one.

        > I know the opposite of any integer is on the opposite side of zero at the same distance. Since −4 is 4 units to the left of zero, then 4 units to the right of zero is 4. The opposite of −4 is 4. The opposite of 8 has to be −8 because −8 is the same distance from zero, just to the left.

2. Write the opposite of each number, and label the points on the number line.

    a. Point A: the opposite of 7   **−7**
    b. Point B: the opposite of −4   **4**
    c. Point C: the opposite of 0   **0**

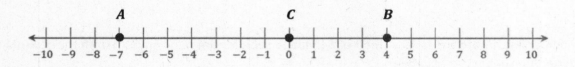

> 7 is located 7 units to the right of zero, so the opposite of 7 must be 7 units to the left of zero. I know −4 is located 4 units to the left of zero, so its opposite has to be 4 units to the right of zero. I also know that zero is its own opposite.

3. Study the first example. Write the integer that represents the opposite of each real-world situation. In words, write the meaning of the opposite.

a. An atom's negative charge of −9

9, *an atom's positive charge of* 9

b. A deposit of $15

−15, *a withdrawal of* $15

c. 2, 500 feet below sea level

2, 500, 2, 500 *feet above sea level*

d. A rise of 35°C

−35, *a decrease of* 35°C

e. A loss of 20 pounds

20, *a gain of* 20 pounds

> I know the following opposites:
> negative/positive
> deposit/withdrawal
> below sea level/above sea level
> rise/decrease
> loss/gain
> Using these opposites, I can determine the opposite of the integers in the situations.

4. On a number line, locate and label a credit of $47 and a debit for the same amount from the bank. What does zero represent in this situation?

Zero represents no change in the balance.

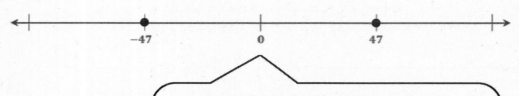

> At the beginning of my transactions, my bank account is a fixed number. If I do not change it, then the change is represented with zero. If I have a credit of 47, I know that that is an increase and falls to the right of zero. If I have a debit of 47, I know that that is a decrease and falls to the left of zero.

1. Find the opposite of each number, and describe its location on the number line.

    a. −5

    b. 10

    c. −3

    d. 15

2. Write the opposite of each number, and label the points on the number line.

    a. Point A: the opposite of 9

    b. Point B: the opposite of −4

    c. Point C: the opposite of −7

    d. Point D: the opposite of 0

    e. Point E: the opposite of 2

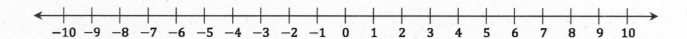

3. Study the first example. Write the integer that represents the opposite of each real-world situation. In words, write the meaning of the opposite.

    a. An atom's positive charge of 7

    b. A deposit of $25

    c. 3,500 feet below sea level

    d. A rise of 45°C

    e. A loss of 13 pounds

4. On a number line, locate and label a credit of $38 and a debit for the same amount from a bank account. What does zero represent in this situation?

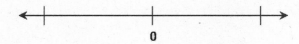

5. On a number line, locate and label 40°C below zero and 40°C above zero. What does zero represent in this situation?

Opening Exercise

a. Locate the number −2 and its opposite on the number line below.

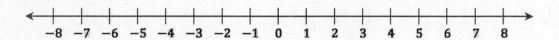

b. Write an integer that represents each of the following.

    i.   90 feet below sea level

    ii. $100 of debt

    iii. 2°C above zero

c. Joe is at the ice cream shop, and his house is 10 blocks north of the shop. The park is 10 blocks south of the ice cream shop. When he is at the ice cream shop, is Joe closer to the park or his house? How could the number zero be used in this situation? Explain.

Example 1: The Opposite of an Opposite of a Number

What is the opposite of the opposite of 8? How can we illustrate this number on a number line?

 a.   What number is 8 units to the right of 0? _____

 b.   How can you illustrate locating the opposite of 8 on this number line?

 c.   What is the opposite of 8? _____

 d.   Use the same process to locate the opposite of −8. What is the opposite of −8? _____

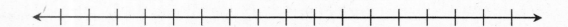

 e.   The opposite of an opposite of a number is _____.

Exercises

Complete the table using the cards in your group.

Person	Card (a)	Opposite of Card ($-a$)	Opposite of Opposite of Card $-(-a)$

1.   Write the opposite of the opposite of −10 as an equation.

2.   In general, the opposite of the opposite of a number is the _____.

3.   Provide a real-world example of this rule. Show your work.

EUREKA
MATH

Name _____   Date _____

1. Jane completes several example problems that ask her to the find the opposite of the opposite of a number, and for each example, the result is a positive number. Jane concludes that when she takes the opposite of the opposite of any number, the result will always be positive. Is Jane correct? Why or why not?

2. To support your answer from the previous question, create an example, written as an equation. Illustrate your example on the number line below.

1. Read each description carefully, and write an equation that represents the description

   a. The opposite of negative six

      $-(-6) = 6$

      > The opposite of a negative number is positive because it is on the opposite side of zero on the number line. The opposite of the opposite of a positive number is positive because the first opposite is on the left side of zero on the number line. The next opposite is to the right of zero.

   b. The opposite of the opposite of thirty-five

      $-(-(35)) = 35$

2. Carol graphed the opposite of the opposite of 4 on the number line. First, she graphed point F on the number line 4 units to the right of zero. Next, she graphed the opposite of F on the number line 4 units to the left of zero and labeled it M. Finally, she graphed the opposite of M and labeled it R.

   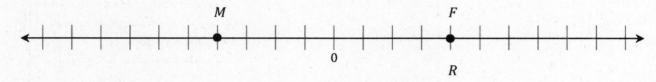

   a. Is her diagram correct? Explain. If the diagram is not correct, explain her error, and correctly locate and label point R.

      Yes, her diagram is correct. It shows that F is 4 because it is 4 units to the right of zero. The opposite of 4 is -4, which is point M (4 units to the left of zero). The opposite of -4 is 4, so point R is 4 units to the right of zero.

   b. Write the relationship between the points.

      F and M

      They are opposites.

      > I see that points M and F are exactly the same distance from zero, just in opposite directions, so they are opposites. M and R are also the same distance from zero on opposite sides, so they are also opposites.

      M and R

      They are opposites.

      F and R

      They are the same.

3. Read each real-world description. Write the integer that represents the opposite of the opposite. Show your work to support your answer.

a. A temperature rise of 20 degrees Fahrenheit

−20 *is the opposite of* 20 *(which is a fall in temperature).*

20 *is the opposite of* −20 *(which is a rise in temperature).*

$-(-(20)) = 20$

> I know that the word *rise* describes a positive integer. The opposite of a positive integer is a negative integer. The opposite of negative integer is a positive integer.

b. A loss of 15 pounds

15 *is the opposite of* −15 *(which is a gain of pounds).*

−15 *is the opposite of* 15 *(which is a loss of pounds).*

$-(-(-15)) = -15$

> I know that the word *loss* describes a negative integer. The opposite of a negative integer is a positive integer. The opposite of a positive integer is a negative integer.

4. Write the integer that represents the statement. Locate and label each integer on the number line below. Plot each integer with a point on the number line.

a. The opposite of a gain of 7   **−7**

b. The opposite of a deposit of $9   **−9**

c. The opposite of the opposite of 0   **0**

d. The opposite of the opposite of 6   **6**

> I know that the words *gain* and *deposit* describe a positive integer.

−9   −7   0   6

Lesson 5:   The Opposite of a Number's Opposite

EUREKA MATH

1. Read each description carefully, and write an equation that represents the description.

    a. The opposite of negative seven

    b. The opposite of the opposite of twenty-five

    c. The opposite of fifteen

    d. The opposite of negative thirty-six

2. Jose graphed the opposite of the opposite of 3 on the number line. First, he graphed point P on the number line 3 units to the right of zero. Next, he graphed the opposite of P on the number line 3 units to the left of zero and labeled it K. Finally, he graphed the opposite of K and labeled it Q.

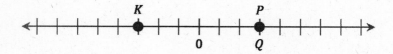

    a. Is his diagram correct? Explain. If the diagram is not correct, explain his error, and correctly locate and label point Q.

    b. Write the relationship between the points:

    P and K _____

    K and Q _____

    P and Q _____

3. Read each real-world description. Write the integer that represents the opposite of the opposite. Show your work to support your answer.

    a. A temperature rise of 15 degrees Fahrenheit

    b. A gain of 55 yards

    c. A loss of 10 pounds

    d. A withdrawal of $2,000

4. Write the integer that represents the statement. Locate and label each point on the number line below.

    a. The opposite of a gain of 6

    b. The opposite of a deposit of $10

    c. The opposite of the opposite of 0

    d. The opposite of the opposite of 4

    e. The opposite of the opposite of a loss of 5

Opening Exercise

a. Write the decimal equivalent of each fraction.

    i. $\dfrac{1}{2}$

    ii. $\dfrac{4}{5}$

    iii. $6\dfrac{7}{10}$

b. Write the fraction equivalent of each decimal.

    i.   0.42

    ii. 3.75

    iii. 36.90

Example 1: Graphing Rational Numbers

If b is a nonzero whole number, then the unit fraction $\frac{1}{b}$ is located on the number line by dividing the segment between 0 and 1 into b segments of equal length. One of the b segments has 0 as its left end point; the right end point of this segment corresponds to the unit fraction $\frac{1}{b}$.

The fraction $\frac{a}{b}$ is located on the number line by joining a segments of length $\frac{1}{b}$ so that (1) the left end point of the first segment is 0, and (2) the right end point of each segment is the left end point of the next segment. The right end point of the last segment corresponds to the fraction $\frac{a}{b}$.

Locate and graph the number $\frac{3}{10}$ and its opposite on a number line.

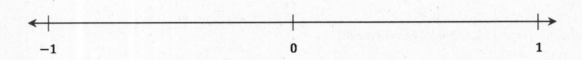

Exercise 1

Use what you know about the point $-\frac{7}{4}$ and its opposite to graph both points on the number line below. The fraction $-\frac{7}{4}$ is located between which two consecutive integers? Explain your reasoning.

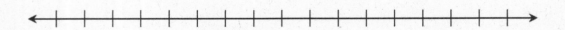

On the number line, each segment will have an equal length of _____. The fraction is located between _____ and _____.

Explanation:

EUREKA
MATH

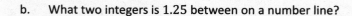

Example 2: Rational Numbers and the Real World

The water level of a lake rose 1.25 feet after it rained. Answer the following questions using the number line below.

a. Write a rational number to represent the situation.

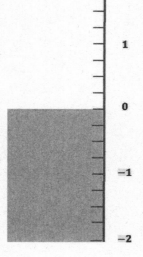

b. What two integers is 1.25 between on a number line?

c. Write the length of each segment on the number line as a decimal and a fraction.

d. What will be the water level after it rained? Graph the point on the number line.

e. After two weeks have passed, the water level of the lake is now the opposite of the water level when it rained. What will be the new water level? Graph the point on the number line. Explain how you determined your answer.

f. State a rational number that is not an integer whose value is less than 1.25, and describe its location between two consecutive integers on the number line.

Exercise 2

Our Story Problem

EUREKA MATH

Name _____   Date _____

Use the number line diagram below to answer the following questions.

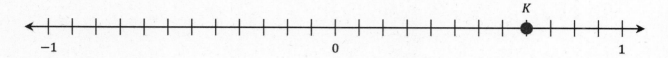

1. What is the length of each segment on the number line?

2. What number does point K represent?

3. What is the opposite of point K?

4. Locate the opposite of point K on the number line, and label it point L.

5. In the diagram above, zero represents the location of Martin Luther King Middle School. Point K represents the library, which is located to the east of the middle school. In words, create a real-world situation that could represent point L, and describe its location in relation to 0 and point K.

1. In the space provided, write the opposite of each number.

   a. $\dfrac{11}{8}$ $-\dfrac{11}{8}$

   b. $-\dfrac{7}{4}$ $\dfrac{7}{4}$

   c. 5.67 -5.67

2. Choose a non-integer between 0 and 1. Label it point A and its opposite B on the number line. Write values below the points.

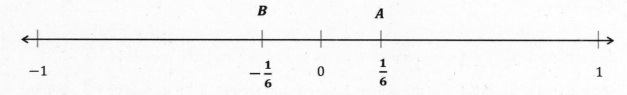

B A

-1 $-\dfrac{1}{6}$ 0 $\dfrac{1}{6}$ 1

> I know integers are numbers that are not fractional. A non-integer is a number that can be a fraction.

   a. To draw a scale that would include both points, what could be the length of each segment?

   The length of each segment could be $\dfrac{1}{6}$.

   b. In words, create a real-world situation that could represent the number line diagram.

   Starting from school, the track is $\dfrac{1}{6}$ of a mile away. The baseball field is $\dfrac{1}{6}$ of a mile from the school in exactly the opposite direction.

3. Choose a value for point P that is between -8 and -9.

    $-\dfrac{26}{3}$

    > I can choose any non-integer less than -8 and more than -9. This can be a fraction or a decimal.

    a. What is the opposite of P?

        $\dfrac{26}{3}$

    b. Use the value from part (a), and describe its location on the number line in relation to zero.

        $\dfrac{26}{3}$ is the same distance as $-\dfrac{26}{3}$ from zero but to the right. $\dfrac{26}{3}$ is $8\dfrac{2}{3}$ units to the right of (or above) zero

    c. Find the opposite of the opposite of point P. Show your work and explain your reasoning.

        The opposite of the opposite of the number is the original number. If P is located at $-\dfrac{26}{3}$, then the opposite of the opposite of P is located at $-\dfrac{26}{3}$. The opposite of $-\dfrac{26}{3}$ is $\dfrac{26}{3}$. The opposite of $\dfrac{26}{3}$ is $-\dfrac{26}{3}$. $-\left(-\left(-\dfrac{26}{3}\right)\right) = -\dfrac{26}{3}$

Lesson 6:   Rational Numbers on the Number Line

EUREKA MATH

4. Locate and label each point on the number line. Use the diagram to answer the questions.

Ami lives one block north of the hair salon.

Trisha's house is $\frac{1}{4}$ of a block past Ami's house.

Isa and Shane are at the soccer field $\frac{6}{4}$ blocks south of the hair salon.

The grocery store is located halfway between the hair salon and the soccer field.

> I know that each of the values in the problem has a denominator of 4, so I separated my number line into equal units of $\frac{1}{4}$. From there, I know one whole is $\frac{4}{4}$ to locate Ami's house.

- Trisha's House
- Ami's House

- 0 Hair Salon

- Grocery Store

- Soccer Field

a. Describe an appropriate scale to show all the points in the situation.

An appropriate scale would be $\frac{1}{4}$ because the numbers given in the example all have denominators of 4. I would divide the number line into equal segments of $\frac{1}{4}$.

b. What number represents the location of the grocery store? Explain your reasoning.

The number is $-\frac{3}{4}$. I found the location of the soccer field, which is 6 units below zero. Half of 6 is 3, so I moved down 3 units from zero.

1. In the space provided, write the opposite of each number.

   a. $\dfrac{10}{7}$

   b. $-\dfrac{5}{3}$

   c. 3.82

   d. $-6\dfrac{1}{2}$

2. Choose a non-integer between 0 and 1. Label it point A and its opposite point B on the number line. Write values below the points.

   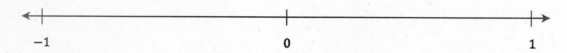

   -1 0 1

   a. To draw a scale that would include both points, what could be the length of each segment?

   b. In words, create a real-world situation that could represent the number line diagram.

3. Choose a value for point P that is between -6 and -7.

   a. What is the opposite of point P?

   b. Use the value from part (a), and describe its location on the number line in relation to zero.

   c. Find the opposite of the opposite of point P. Show your work, and explain your reasoning.

4. Locate and label each point on the number line. Use the diagram to answer the questions.

   Jill lives one block north of the pizza shop.

   Janette's house is $\dfrac{1}{3}$ block past Jill's house.

   Jeffrey and Olivia are in the park $\dfrac{4}{3}$ blocks south of the pizza shop.

   Jenny's Jazzy Jewelry Shop is located halfway between the pizza shop and the park.

   a. Describe an appropriate scale to show all the points in this situation.

   b. What number represents the location of Jenny's Jazzy Jewelry Shop? Explain your reasoning.

Exercise 1

    a.   Graph the number 7 and its opposite on the number line. Graph the number 5 and its opposite on the number line.

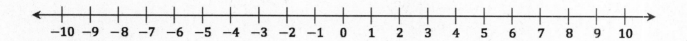

    b.   Where does 7 lie in relation to 5 on the number line?

    c.   Where does the opposite of 7 lie on the number line in relation to the opposite of 5?

    d.   I am thinking of two numbers. The first number lies to the right of the second number on a number line. What can you say about the location of their opposites? (If needed, refer to your number line diagram.)

Example 1

The record low temperatures for a town in Maine are $-20°F$ for January and $-19°F$ for February. Order the numbers from least to greatest. Explain how you arrived at the order.

Exercises 2–4

For each problem, order the rational numbers from least to greatest by first reading the problem, then drawing a number line diagram, and finally, explaining your answer.

2. Jon's time for running the mile in gym class is 9.2 minutes. Jacky's time is 9.18 minutes. Who ran the mile in less time?

3. Mrs. Rodriguez is a teacher at Westbury Middle School. She gives bonus points on tests for outstanding written answers and deducts points for answers that are not written correctly. She uses rational numbers to represent the points. She wrote the following on students' papers: Student A: -2 points, Student B: -2.5 points. Did Student A or Student B perform worse on the test?

4. A carp is swimming approximately $8\frac{1}{4}$ feet beneath the water's surface, and a sunfish is swimming approximately $3\frac{1}{2}$ feet beneath the water's surface. Which fish is swimming farther beneath the water's surface?

Example 2

Henry, Janon, and Clark are playing a card game. The object of the game is to finish with the most points. The scores at the end of the game are Henry: -7, Janon: 0, and Clark: -5. Who won the game? Who came in last place? Use a number line model, and explain how you arrived at your answer.

Lesson 7:   Ordering Integers and Other Rational Numbers

© 2019 Great Minds®. eureka-math.org

EUREKA MATH

Exercises 5–6

For each problem, order the rational numbers from least to greatest by first reading the problem, then drawing a number line diagram, and finally, explaining your answer.

5. Henry, Janon, and Clark are playing another round of the card game. Their scores this time are as follows: Clark: −1, Janon: −2, and Henry: −4. Who won? Who came in last place?

6. Represent each of the following elevations using a rational number. Then, order the numbers from least to greatest.

Cayuga Lake   122 meters above sea level

Mount Marcy   1,629 meters above sea level

New York Stock Exchange Vault   15.24 meters below sea level

Closing: What Is the Value of Each Number, and Which Is Larger?

Use your teacher's verbal clues and this number line to determine which number is larger.

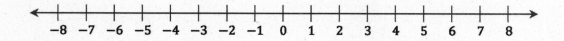

Name _____   Date _____

In math class, Christina and Brett are debating the relationship between two rational numbers. Read their claims below, and then write an explanation of who is correct. Use a number line model to support your answer.

<u>Christina's Claim</u>: "I know that 3 is greater than $2\frac{1}{2}$. So, -3 must be greater than $-2\frac{1}{2}$."

<u>Brett's Claim</u>: "Yes, 3 is greater than $2\frac{1}{2}$, but when you look at their opposites, their order will be opposite. So, that means $-2\frac{1}{2}$ is greater than -3."

1. In the table below, list each set of rational numbers in order from least to greatest. Then, list their opposites. Finally, list the opposites in order from least to greatest.

Rational Numbers	Ordered from Least to Greatest	Opposites	Opposites Ordered from Least to Greatest
$-6.1, -6.35$	$-6.35, -6.1$	$6.35, 6.1$	$6.1, 6.35$
$\frac{1}{3}, -\frac{1}{4}$	$-\frac{1}{4}, \frac{1}{3}$	$\frac{1}{4}, -\frac{1}{3}$	$-\frac{1}{3}, \frac{1}{4}$
$-49.9, -50$	$-50, -49.9$	$50, 49.9$	$49.9, 50$
$32\frac{1}{3}, 32$	$32, 32\frac{1}{3}$	$-32, -32\frac{1}{3}$	$-32\frac{1}{3}, -32$
$65.03, 65.05$	$65.03, 65.05$	$-65.03, -65.05$	$-65.05, -65.03$

> I can visualize a number line to order the rational numbers from least to greatest. The number farthest to the left on the number line is the least, and the number farthest to the right is the greatest.

2. For each row, what pattern do you notice between the numbers in the second and fourth columns? Why is this so?

For each row, the numbers in the second and fourth columns are opposites, and their order is opposite. This is because on the number line, as you move to the right, numbers increase. But as you move to the left, numbers decrease.

1. In the table below, list each set of rational numbers in order from least to greatest. Then, list their opposites. Finally, list the opposites in order from least to greatest. The first example has been completed for you.

Rational Numbers	Ordered from Least to Greatest	Opposites	Opposites Ordered from Least to Greatest
$-7.1, -7.25$	$-7.25, -7.1$	$7.25, 7.1$	$7.1, 7.25$
$\frac{1}{4}, -\frac{1}{2}$			
$2, -10$			
$0, 3\frac{1}{2}$			
$-5, -5.6$			
$24\frac{1}{2}, 24$			
$-99.9, -100$			
$-0.05, -0.5$			
$-0.7, 0$			
$100.02, 100.04$			

2. For each row, what pattern do you notice between the numbers in the second and fourth columns? Why is this so?

Example 1: Ordering Rational Numbers from Least to Greatest

Sam has $10.00 in the bank. He owes his friend Hank $2.25. He owes his sister $1.75. Consider the three rational numbers related to this story of Sam's money. Write and order them from least to greatest.

Exercises 2–4

For each problem, list the rational numbers that relate to each situation. Then, order them from least to greatest, and explain how you made your determination.

2. During their most recent visit to the optometrist (eye doctor), Kadijsha and her sister, Beth, had their vision tested. Kadijsha's vision in her left eye was -1.50, and her vision in her right eye was the opposite number. Beth's vision was -1.00 in her left eye and $+0.25$ in her right eye.

3. There are three pieces of mail in Ms. Thomas's mailbox: a bill from the phone company for $38.12, a bill from the electric company for $67.55, and a tax refund check for $25.89. (A bill is money that you owe, and a tax refund check is money that you receive.)

4. Monica, Jack, and Destiny measured their arm lengths for an experiment in science class. They compared their arm lengths to a standard length of 22 inches. The listing below shows, in inches, how each student's arm length compares to 22 inches.

Monica: $-\dfrac{1}{8}$

Jack: $1\dfrac{3}{4}$

Destiny: $-\dfrac{1}{2}$

Example 2: Ordering Rational Numbers from Greatest to Least

Jason is entering college and has opened a checking account, which he will use for college expenses. His parents gave him $200.00 to deposit into the account. Jason wrote a check for $85.00 to pay for his calculus book and a check for $25.34 to pay for miscellaneous school supplies. Write the three rational numbers related to the balance in Jason's checking account in order from greatest to least.

Exercises 5–6

For each problem, list the rational numbers that relate to each situation in order from greatest to least. Explain how you arrived at the order.

5. The following are the current monthly bills that Mr. McGraw must pay:

   $122.00 Cable and Internet

   $73.45 Gas and Electric

   $45.00 Cell Phone

6. $-\dfrac{1}{3}, 0, -\dfrac{1}{5}, \dfrac{1}{8}$

EUREKA MATH

Lesson Summary

When we order rational numbers, their opposites are in the opposite order. For example, if 7 is greater than 5, −7 is less than −5.

Name _____ Date _____

Order the following set of rational numbers from least to greatest, and explain how you determined the order.

$$-3, 0, -\frac{1}{2}, 1, -3\frac{1}{3}, 6, 5, -1, \frac{21}{5}, 4$$

1. In the table below, list each set of rational numbers in order from greatest to least. Then, in the appropriate column, state which number was farthest right and which number was farthest left on the number line.

Column 1	Column 2	Column 3	Column 4
Rational Numbers	Ordered from Greatest to Least	Farthest Right on the Number Line	Farthest Left on the Number Line
$-2.85, -4.15$	$-2.85, -4.15$	-2.85	-4.15
$\frac{1}{3}, -3$	$\frac{1}{3}, -3$	$\frac{1}{3}$	-3
$0.04, 0.4$	$0.4, 0.04$	0.4	0.04
$0, -\frac{1}{3}, -\frac{2}{3}$	$0, -\frac{1}{3}, -\frac{2}{3}$	0	$-\frac{2}{3}$

> I can visualize a number line to order the rational numbers from greatest to least. The number farthest to the right on the number line is the greatest. The number farthest to the left is the least number.

a. For each row, describe the relationship between the number in Column 3 and its order in Column 2. Why is this?

The number in Column 3 is the first number listed in Column 2. Since it is the farthest right number on the number line, it will be the greatest; therefore, it comes first when ordering the numbers from greatest to least.

b. For each row, describe the relationship between the number in Column 4 and its order in Column 2. Why is this?

The number in Column 4 is the last number listed in Column 2. Since it is farthest left on the number line, it will be the least; therefore, it comes last when ordering from greatest to least.

2. If two rational numbers, a and b, are ordered such that a is less than b, then what must be true about the order of their opposites: $-a$ and $-b$?

   The order will be reversed for the opposites, which means $-a$ is greater than $-b$.

3. Read each statement, and then write a statement relating the *opposites* of each of the given numbers.

   a. 8 is greater than 7.

      -8 *is less than* -7.

   b. 48.1 is greater than 40.

      -48.1 *is less than* -40.

      > I notice that the order is reversed for the opposites.

   c. $-\dfrac{1}{2}$ is less than $-\dfrac{1}{6}$.

      $\dfrac{1}{2}$ *is greater than* $\dfrac{1}{6}$.

4. Order the following from least to greatest: $-8, -17, 0, \dfrac{1}{2}, \dfrac{1}{4}$.

   $-17, -8, 0, \dfrac{1}{4}, \dfrac{1}{2}$

   > When I order from least to greatest, I think about the number that is farthest left on the number line. When I order from greatest to least, I start with the number farthest to the right on the number line.

5. Order the following from greatest to least: $-14, 14, -20, 2\dfrac{1}{2}, 7$.

   $14, 7, 2\dfrac{1}{2}, -14, -20$

Lesson 8:   Ordering Integers and Other Rational Numbers

© 2019 Great Minds®. eureka-math.org

EUREKA MATH

1.

a. In the table below, list each set of rational numbers from greatest to least. Then, in the appropriate column, state which number was farthest right and which number was farthest left on the number line.

Column 1	Column 2	Column 3	Column 4
Rational Numbers	Ordered from Greatest to Least	Farthest Right on the Number Line	Farthest Left on the Number Line
$-1.75, -3.25$			
$-9.7, -9$			
$\frac{4}{5}, 0$			
$-70, -70\frac{4}{5}$			
$-15, -5$			
$\frac{1}{2}, -2$			
$-99, -100, -99.3$			
$0.05, 0.5$			
$0, -\frac{3}{4}, -\frac{1}{4}$			
$-0.02, -0.04$			

b. For each row, describe the relationship between the number in Column 3 and its order in Column 2. Why is this?

c. For each row, describe the relationship between the number in Column 4 and its order in Column 2. Why is this?

2. If two rational numbers, a and b, are ordered such that a is less than b, then what must be true about the order for their opposites: $-a$ and $-b$?

3. Read each statement, and then write a statement relating the *opposites* of each of the given numbers:

a. 7 is greater than 6.

b. 39.2 is greater than 30.

c. $-\frac{1}{5}$ is less than $\frac{1}{3}$.

4. Order the following from least to greatest: $-8, -19, 0, \frac{1}{2}, \frac{1}{4}$.

5. Order the following from greatest to least: $-12, 12, -19, 1\frac{1}{2}, 5$.

Lesson 8:   Ordering Integers and Other Rational Numbers

EUREKA MATH

Example 1: Interpreting Number Line Models to Compare Numbers

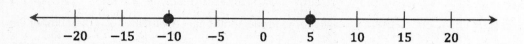

Exercises

1. Create a real-world situation that relates to the points shown in the number line model. Be sure to describe the relationship between the values of the two points and how it relates to their order on the number line.

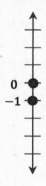

For each problem, determine if you *agree or disagree* with the representation. Then, defend your stance by citing specific details in your writing.

2. Felicia needs to write a story problem that relates to the order in which the numbers $-6\frac{1}{2}$ and -10 are represented on a number line. She writes the following:

    "During a recent football game, our team lost yards on two consecutive downs. We lost $6\frac{1}{2}$ yards on the first down. During the second down, our quarterback was sacked for an additional 10-yard loss. On the number line, I represented this situation by first locating $-6\frac{1}{2}$. I located the point by moving $6\frac{1}{2}$ units to the left of zero. Then, I graphed the second point by moving 10 units to the left of 0."

3. Manuel looks at a number line diagram that has the points $-\frac{3}{4}$ and $-\frac{1}{2}$ graphed. He writes the following related story:

    "I borrowed 50 cents from my friend, Lester. I borrowed 75 cents from my friend, Calvin. I owe Lester less than I owe Calvin."

4. Henry located $2\frac{1}{4}$ and 2.1 on a number line. He wrote the following related story:

    "In gym class, both Jerry and I ran for 20 minutes. Jerry ran $2\frac{1}{4}$ miles, and I ran 2.1 miles. I ran a farther distance."

EUREKA MATH

5. Sam looked at two points that were graphed on a vertical number line. He saw the points -2 and 1.5. He wrote the following description:

"I am looking at a vertical number line that shows the location of two specific points. The first point is a negative number, so it is below zero. The second point is a positive number, so it is above zero. The negative number is -2. The positive number is $\frac{1}{2}$ unit more than the negative number."

6. Claire draws a vertical number line diagram and graphs two points: -10 and 10. She writes the following related story:

"These two locations represent different elevations. One location is 10 feet above sea level, and one location is 10 feet below sea level. On a number line, 10 feet above sea level is represented by graphing a point at 10, and 10 feet below sea level is represented by graphing a point at -10."

7. Mrs. Kimble, the sixth-grade math teacher, asked the class to describe the relationship between two points on the number line, 7.45 and 7.5, and to create a real-world scenario. Jackson writes the following story:

"Two friends, Jackie and Jennie, each brought money to the fair. Jackie brought more than Jennie. Jackie brought $7.45, and Jennie brought $7.50. Since 7.45 has more digits than 7.5, it would come after 7.5 on the number line, or to the right, so it is a greater value."

8. Justine graphs the points associated with the following $\frac{1}{2}$ numbers on a vertical number line: $-1\frac{1}{4}, -1\frac{1}{2}$, and 1. She then writes the following real-world scenario:

"The nurse measured the height of three sixth-grade students and compared their heights to the height of a typical sixth grader. Two of the students' heights are below the typical height, and one is above the typical height. The point whose coordinate is 1 represents the student who has a height that is 1 inch above the typical height. Given this information, Justine determined that the student represented by the point associated with $-1\frac{1}{4}$ is the shortest of the three students."

EUREKA MATH

Name _____   Date _____

1. Interpret the number line diagram shown below, and write a statement about the temperature for Tuesday compared to Monday at 11:00 p.m.

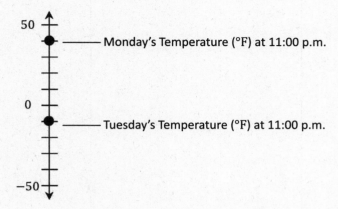

2. If the temperature at 11:00 p.m. on Wednesday is warmer than Tuesday's temperature but still below zero, what is a possible value for the temperature at 11:00 p.m. Wednesday?

Write a story related to the points shown in each group. Be sure to include a statement relating the numbers graphed on the number line to their order.

1.

   Julia did not improve on her Sprint yesterday. Today, she improved her score by three points. Zero represents earning no improvement points yesterday, and 3 represents earning 3 improvement points. Zero is graphed to the left of 3 on the number line. Zero is less than 3.

2.

   A turtle is swimming one foot below the surface of the water. An eel is swimming $8\frac{1}{2}$ feet below the water's surface. $-8\frac{1}{2}$ is farther below zero than -1, so the eel is swimming deeper than the turtle.

   I know that as numbers are farther down a vertical number line, the values of the numbers decrease. The greater of two numbers is the number that is farthest up.

Write a story related to the points shown in each graph. Be sure to include a statement relating the numbers graphed on the number line to their order.

1.

2.

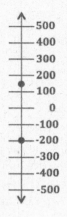

3.

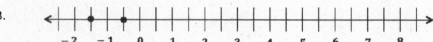

4.

5.

6.

7.

Number Correct: _____

Rational Numbers: Inequality Statements—Round 1

Directions: Work in numerical order to answer Problems 1–33. Arrange each set of numbers in order according to the inequality symbols.

1. ☐ < ☐ < ☐ $1, -1, 0$	**12.** ☐ > ☐ > ☐ $7, -6, 6$	**23.** ☐ > ☐ > ☐ $25, \frac{3}{4}, -\frac{3}{4}$
2. ☐ > ☐ > ☐ $1, -1, 0$	**13.** ☐ > ☐ > ☐ $17, 4, 16$	**24.** ☐ < ☐ < ☐ $25, \frac{3}{4}, -\frac{3}{4}$
3. ☐ < ☐ < ☐ $3\frac{1}{2}, -3\frac{1}{2}, 0$	**14.** ☐ < ☐ < ☐ $17, 4, 16$	**25.** ☐ > ☐ > ☐ $2.2, 2.3, 2.4$
4. ☐ > ☐ > ☐ $3\frac{1}{2}, -3\frac{1}{2}, 0$	**15.** ☐ < ☐ < ☐ $0, 12, -11$	**26.** ☐ > ☐ > ☐ $1.2, 1.3, 1.4$
5. ☐ > ☐ > ☐ $1, -\frac{1}{2}, \frac{1}{2}$	**16.** ☐ > ☐ > ☐ $0, 12, -11$	**27.** ☐ > ☐ > ☐ $0.2, 0.3, 0.4$
6. ☐ < ☐ < ☐ $1, -\frac{1}{2}, \frac{1}{2}$	**17.** ☐ > ☐ > ☐ $1, \frac{1}{4}, \frac{1}{2}$	**28.** ☐ > ☐ > ☐ $-0.5, -1, -0.6$
7. ☐ < ☐ < ☐ $-3, -4, -5$	**18.** ☐ < ☐ < ☐ $1, \frac{1}{4}, \frac{1}{2}$	**29.** ☐ < ☐ < ☐ $-0.5, -1, -0.6$
8. ☐ < ☐ < ☐ $-13, -14, -15$	**19.** ☐ < ☐ < ☐ $-\frac{1}{2}, \frac{1}{2}, 0$	**30.** ☐ < ☐ < ☐ $-8, -9, 8$
9. ☐ > ☐ > ☐ $-13, -14, -15$	**20.** ☐ > ☐ > ☐ $-\frac{1}{2}, \frac{1}{2}, 0$	**31.** ☐ < ☐ < ☐ $-18, -19, -2$
10. ☐ < ☐ < ☐ $-\frac{1}{4}, -1, 0$	**21.** ☐ < ☐ < ☐ $50, -10, 0$	**32.** ☐ > ☐ > ☐ $-2, -3, 1$
11. ☐ > ☐ > ☐ $-\frac{1}{4}, -1, 0$	**22.** ☐ < ☐ < ☐ $-50, 10, 0$	**33.** ☐ < ☐ < ☐ $-2, -3, 1$

Number Correct: _____

Improvement: _____

Rational Numbers: Inequality Statements—Round 2

Directions: Work in numerical order to answer Problems 1–33. Arrange each set of numbers in order according to the inequality symbols.

1. ☐ < ☐ < ☐
1/7 , −1/7 , 0

2. ☐ > ☐ > ☐
1/7 , −1/7 , 0

3. ☐ < ☐ < ☐
3/7 , 2/7 , −1/7

4. ☐ > ☐ > ☐
3/7 , 2/7 , −1/7

5. ☐ > ☐ > ☐
−4/5 , 1/5 , −1/5

6. ☐ < ☐ < ☐
−4/5 , 1/5 , −1/5

7. ☐ < ☐ < ☐
−8/9 , 5/9 , 1/9

8. ☐ > ☐ > ☐
−8/9 , 5/9 , 1/9

9. ☐ > ☐ > ☐
−30 , −10 , −50

10. ☐ < ☐ < ☐
−30 , −10 , −50

11. ☐ > ☐ > ☐
−40 , −20 , −60

12. ☐ > ☐ > ☐
1¼ , 1 , 1½

13. ☐ > ☐ > ☐
11¼ , 11 , 11½

14. ☐ < ☐ < ☐
11¼ , 11 , 11½

15. ☐ < ☐ < ☐
0 , 0.2 , −0.1

16. ☐ > ☐ > ☐
0 , 0.2 , −0.1

17. ☐ > ☐ > ☐
1 , 0.7 , 1/10

18. ☐ < ☐ < ☐
1 , 0.7 , 1/10

19. ☐ < ☐ < ☐
0 , −12 , −12½

20. ☐ > ☐ > ☐
0 , −12 , −12½

21. ☐ < ☐ < ☐
5 , −1 , 0

22. ☐ < ☐ < ☐
−5 , 1 , 0

23. ☐ > ☐ > ☐
1 , 1¾ , − 1¾

24. ☐ < ☐ < ☐
1 , 1¾ , − 1¾

25. ☐ > ☐ > ☐
−82 , −93 , −104

26. ☐ < ☐ < ☐
−82 , −93 , −104

27. ☐ > ☐ > ☐
0.5 , 1 , 0.6

28. ☐ > ☐ > ☐
−0.5 , − 1 , −0.6

29. ☐ < ☐ < ☐
−0.5 , − 1 , −0.6

30. ☐ < ☐ < ☐
1 , 8 , 9

31. ☐ < ☐ < ☐
−1 , −8 , −9

32. ☐ > ☐ > ☐
−2 , −3 , −5

33. ☐ > ☐ > ☐
2 , 3 , 5

EUREKA MATH®

Opening Exercise

"The amount of money I have in my pocket is less than $5 but greater than $4."

    a.   One possible value for the amount of money in my pocket is _____.

    b.   Write an inequality statement comparing the possible value of the money in my pocket to $4.

    c.   Write an inequality statement comparing the possible value of the money in my pocket to $5.

Exercises 1–4

1.   Graph your answer from the Opening Exercise part (a) on the number line below.

2.   Also, graph the points associated with 4 and 5 on the number line.

3.   Explain in words how the location of the three numbers on the number line supports the inequality statements you wrote in the Opening Exercise parts (b) and (c).

4.   Write one inequality statement that shows the relationship among all three numbers.

Example 1: Writing Inequality Statements Involving Rational Numbers

Write one inequality statement to show the relationship among the following shoe sizes: $10\frac{1}{2}$, 8, and 9.

    a.   From least to greatest:

    b.   From greatest to least:

Example 2: Interpreting Data and Writing Inequality Statements

Mary is comparing the rainfall totals for May, June, and July. The data is reflected in the table below. Fill in the blanks below to create inequality statements that compare the Changes in Total Rainfall for each month (the right-most column of the table).

Month	This Year's Total Rainfall (in inches)	Last Year's Total Rainfall (in inches)	Change in Total Rainfall from Last Year to This Year (in inches)
May	2.3	3.7	−1.4
June	3.8	3.5	0.3
July	3.7	3.2	0.5

Write one inequality to order the Changes in Total Rainfall: _____ _____

                                        From least to greatest   From greatest to least

In this case, does the greatest number indicate the greatest change in rainfall? Explain.

EUREKA MATH

Exercises 5–8

5. Mark's favorite football team lost yards on two back-to-back plays. They lost 3 yards on the first play. They lost 1 yard on the second play. Write an inequality statement using integers to compare the forward progress made on each play.

6. Sierra had to pay the school for two textbooks that she lost. One textbook cost $55, and the other cost $75. Her mother wrote two separate checks for each expense. Write two integers that represent the change to her mother's checking account balance. Then, write an inequality statement that shows the relationship between these two numbers.

7. Jason ordered the numbers -70, -18, and -18.5 from least to greatest by writing the following statement: $-18 < -18.5 < -70$.
    Is this a true statement? Explain.

8. Write a real-world situation that is represented by the following inequality: $-19 < 40$. Explain the position of the numbers on a number line.

Exercise 9: A Closer Look at the Sprint

9. Look at the following two examples from the Sprint.

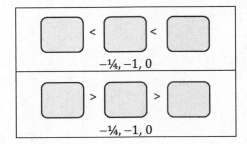

a. Fill in the numbers in the correct order.

b. Explain how the position of the numbers on the number line supports the inequality statements you created.

c. Create a new pair of greater than and less than inequality statements using three other rational numbers.

         Lesson 10:   Writing and Interpreting Inequality Statements
                                   Involving Rational Numbers

EUREKA MATH

Name _____   Date _____

Kendra collected data for her science project. She surveyed people asking them how many hours they sleep during a typical night. The chart below shows how each person's response compares to 8 hours (which is the answer she expected most people to say).

Name	Number of Hours (usually slept each night)	Compared to 8 Hours
Frankie	8.5	0.5
Mr. Fields	7	-1.0
Karla	9.5	1.5
Louis	8	0
Tiffany	$7\frac{3}{4}$	$-\frac{1}{4}$

a.   Plot and label each of the numbers in the right-most column of the table above on the number line below.

b.   List the numbers from least to greatest.

c.   Using your answer from part (b) and inequality symbols, write one statement that shows the relationship among all of the numbers.

For each of the relationships described below, write an inequality that relates the rational numbers.

1. Ten feet below sea level is farther below sea level than $5\frac{1}{4}$ feet below sea level.

$$-10 < -5\frac{1}{4}$$

2. Kelly's grades on her last three tests were 85, 90, and $75\frac{1}{2}$. A score of $75\frac{1}{2}$ is worse than a score of 85. A score of 85 is worse than a score of 90.

$$75\frac{1}{2} < 85 < 90$$

For each of the following, use the information given by the inequality to describe the relative position of the numbers on a horizontal number line.

3. $-3.4 < 0 < 3.2$

   −3.4 is to the left of zero, and zero is to the left of 3.2; or 3.2 is to the right of zero, and zero is to the right of −3.4.

4. $-5.7 < -5\frac{1}{2} < -5$

   −5.7 is to the left of $-5\frac{1}{2}$, and $-5\frac{1}{2}$ is to the left of −5; or −5 is to the right of $-5\frac{1}{2}$, and $-5\frac{1}{2}$ is to the right of −5.7.

Fill in the blanks with numbers that correctly complete each of the statements.

5. Three integers between −5 and −1   −4, −3, −2

6. Three rational numbers between −3 and −4   −3.45, −3.6, −3.99

> Any rational number between −3 and −4 is acceptable.

For each of the relationships described below, write an inequality that relates the rational numbers.

1. Seven feet below sea level is farther below sea level than $4\frac{1}{2}$ feet below sea level.

2. Sixteen degrees Celsius is warmer than zero degrees Celsius.

3. Three and one-half yards of fabric is less than five and one-half yards of fabric.

4. A loss of $500 in the stock market is worse than a gain of $200 in the stock market.

5. A test score of 64 is worse than a test score of 65, and a test score of 65 is worse than a test score of $67\frac{1}{2}$.

6. In December, the total snowfall was 13.2 inches, which is more than the total snowfall in October and November, which was 3.7 inches and 6.15 inches, respectively.

For each of the following, use the information given by the inequality to describe the relative position of the numbers on a horizontal number line.

7. $-0.2 < -0.1$

8. $8\frac{1}{4} > -8\frac{1}{4}$

9. $-2 < 0 < 5$

10. $-99 > -100$

11. $-7.6 < -7\frac{1}{2} < -7$

Fill in the blanks with numbers that correctly complete each of the statements.

12. Three integers between -4 and 0 _____ < _____ < _____

13. Three rational numbers between 16 and 15   _____ < _____ < _____

14. Three rational numbers between -1 and -2 _____ < _____ < _____

15. Three integers between 2 and -2 _____ < _____ < _____

EUREKA
MATH

Lesson 10:   Writing and Interpreting Inequality Statements
              Involving Rational Numbers

103

© 2019 Great Minds®. eureka-math.org

Opening Exercise

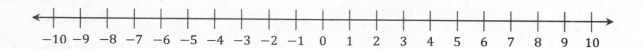

Example 1: The Absolute Value of a Number

The absolute value of ten is written as |10|. On the number line, count the number of units from 10 to 0. How many units is 10 from 0?

|10| =

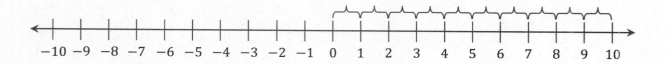

What other number has an absolute value of 10? Why?

The <u>absolute value</u> of a number is the distance between the number and zero on the number line.

Exercises 1–3

Complete the following chart.

	Number	Absolute Value	Number Line Diagram	Different Number with the Same Absolute Value
1.	−6			
2.	8			
3.	−1			

Example 2: Using Absolute Value to Find Magnitude

Mrs. Owens received a call from her bank because she had a checkbook balance of −$45. What was the magnitude of the amount overdrawn?

The <u>magnitude</u> of a measurement is the absolute value of its measure.

Exercises 4–19

For each scenario below, use absolute value to determine the magnitude of each quantity.

4. Maria was sick with the flu, and her weight change as a result of it is represented by −4 pounds. How much weight did Maria lose?

5. Jeffrey owes his friend $5. How much is Jeffrey's debt?

6. The elevation of Niagara Falls, which is located between Lake Erie and Lake Ontario, is 326 feet. How far is this above sea level?

7. How far below zero is −16 degrees Celsius?

8. Frank received a monthly statement for his college savings account. It listed a deposit of $100 as +100.00. It listed a withdrawal of $25 as −25.00. The statement showed an overall ending balance of $835.50. How much money did Frank add to his account that month? How much did he take out? What is the total amount Frank has saved for college?

9. Meg is playing a card game with her friend, Iona. The cards have positive and negative numbers printed on them. Meg exclaims: "The absolute value of the number on my card equals 8." What is the number on Meg's card?

10. List a positive and negative number whose absolute value is greater than 3. Justify your answer using the number line.

11. Which of the following situations can be represented by the absolute value of 10? Check all that apply.

____ The temperature is 10 degrees below zero. Express this as an integer.

____ Determine the size of Harold's debt if he owes $10.

____ Determine how far −10 is from zero on a number line.

____ 10 degrees is how many degrees above zero?

12. Julia used absolute value to find the distance between 0 and 6 on a number line. She then wrote a similar statement to represent the distance between 0 and −6. Below is her work. Is it correct? Explain.

$$|6| = 6 \text{ and } |-6| = -6$$

13. Use absolute value to represent the amount, in dollars, of a $238.25 profit.

14. Judy lost 15 pounds. Use absolute value to represent the number of pounds Judy lost.

15. In math class, Carl and Angela are debating about integers and absolute value. Carl said two integers can have the same absolute value, and Angela said one integer can have two absolute values. Who is right? Defend your answer.

16. Jamie told his math teacher: "Give me any absolute value, and I can tell you two numbers that have that absolute value." Is Jamie correct? For any given absolute value, will there always be two numbers that have that absolute value?

17. Use a number line to show why a number and its opposite have the same absolute value.

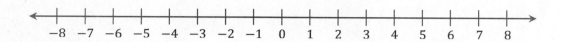

18. A bank teller assisted two customers with transactions. One customer made a $25 withdrawal from a savings account. The other customer made a $15 deposit. Use absolute value to show the size of each transaction. Which transaction involved more money?

19. Which is farther from zero: $-7\frac{3}{4}$ or $7\frac{1}{2}$? Use absolute value to defend your answer.

Name _____   Date _____

Jessie and his family drove up to a picnic area on a mountain. In the morning, they followed a trail that led to the mountain summit, which was 2,000 feet above the picnic area. They then returned to the picnic area for lunch. After lunch, they hiked on a trail that led to the mountain overlook, which was 3,500 feet below the picnic area.

a. Locate and label the elevation of the mountain summit and mountain overlook on a vertical number line. The picnic area represents zero. Write a rational number to represent each location.

   Picnic area: ___0___

   Mountain summit: _____

   Mountain overlook: _____

b. Use absolute value to represent the distance on the number line of each location from the picnic area.

   Distance from the picnic area to the mountain summit: _____

   Distance from the picnic area to the mountain overlook: _____

c. What is the distance between the elevations of the summit and overlook? Use absolute value and your number line from part (a) to explain your answer.

1. For the following two quantities, which has the greater magnitude? (Use absolute value to defend your answers.)

    −13.6 pounds and −13.68 pounds

    $|-13.6| = 13.6$ $|-13.68| = 13.68$

    $13.6 < 13.68$, so -13.68 has the greater magnitude.

    > I can find the absolute value of both numbers and compare. The *magnitude* of a measurement is the absolute value of its measure.

2. Find the absolute value of the numbers below.

    a. $|8| =$

    b. $|-96.2| =$

    c. $|0| =$

    > In part (a), 8 is 8 units from 0, so the absolute value of 8 is 8. -96.2 is 96.2 units from 0, so its absolute value is 96.2. The absolute value of 0 is 0 and is neither positive nor negative.

    a. $|8| = 8$

    b. $|-96.2| = 96.2$

    c. $|0| = 0$

3. Write a word problem whose solution is $|150| = 150$.

    Answers will vary. Kendra went hiking and was 150 feet above sea level.

    > If sea level is the reference point, I know a positive number (150) will represent a number above sea level, and a negative number (−80) will represent a number below sea level.

4. Write a word problem whose solution is $|-80| = 80$.

    Answers will vary. Kristen went scuba diving and was 80 feet below sea level.

For each of the following two quantities in Problems 1–4, which has the greater magnitude? (Use absolute value to defend your answers.)

1. 33 dollars and −52 dollars

2. −14 feet and 23 feet

3. −24.6 pounds and −24.58 pounds

4. $-11\frac{1}{4}$ degrees and 11 degrees

For Problems 5–7, answer true or false. If false, explain why.

5. The absolute value of a negative number will always be a positive number.

6. The absolute value of any number will always be a positive number.

7. Positive numbers will always have a higher absolute value than negative numbers.

8. Write a word problem whose solution is $|20| = 20$.

9. Write a word problem whose solution is $|-70| = 70$.

10. Look at the bank account transactions listed below, and determine which has the greatest impact on the account balance. Explain.
    a. A withdrawal of $60
    b. A deposit of $55
    c. A withdrawal of $58.50

Opening Exercise

Record your integer values in order from least to greatest in the space below.

Write an inequality statement relating the ordered integers from the Opening Exercise. Below each integer, write its absolute value.

Circle the absolute values that are in increasing numerical order and their corresponding integers. Describe the circled values.

Rewrite the integers that are not circled in the space below. How do these integers differ from the ones you circled?

Rewrite the negative integers in ascending order and their absolute values in ascending order below them.

Describe how the order of the absolute values compares to the order of the negative integers.

Example 2: The Order of Negative Integers and Their Absolute Values

Draw arrows starting at the dashed line (zero) to represent each of the integers shown on the number line below. The arrows that correspond with 1 and 2 have been modeled for you.

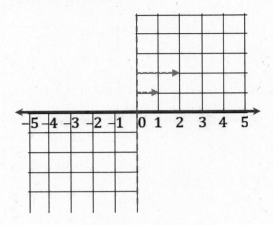

As you approach zero from the left on the number line, the integers _____, but the absolute values of those integers _____. This means that the order of negative integers is _____ the order of their absolute values.

Exercise 1

Complete the steps below to order these numbers:

$$\left\{2.1, -4\frac{1}{2}, -6, 0.25, -1.5, 0, 3.9, -6.3, -4, 2\frac{3}{4}, 3.99, -9\frac{1}{4}\right\}$$

a. Separate the set of numbers into positive rational numbers, negative rational numbers, and zero in the top cells below (order does not matter).

b. Write the absolute values of the rational numbers (order does not matter) in the bottom cells below.

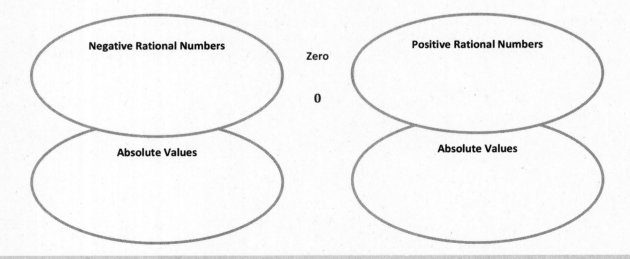

c. Order each subset of absolute values from least to greatest.

[ ] 0 [ ]

d. Order each subset of rational numbers from least to greatest.

[ ] 0 [ ]

e. Order the whole given set of rational numbers from least to greatest.

[ ]

Exercise 2

a. Find a set of four integers such that their order and the order of their absolute values are the same.

b. Find a set of four integers such that their order and the order of their absolute values are opposite.

c. Find a set of four non-integer rational numbers such that their order and the order of their absolute values are the same.

d. Find a set of four non-integer rational numbers such that their order and the order of their absolute values are opposite.

e. Order all of your numbers from parts (a)–(d) in the space below. This means you should be ordering 16 numbers from least to greatest.

> ### Lesson Summary
>
> The absolute values of positive numbers always have the same order as the positive numbers themselves. Negative numbers, however, have exactly the opposite order as their absolute values. The absolute values of numbers on the number line increase as you move away from zero in either direction.

Lesson 12: The Relationship Between Absolute Value and Order

EUREKA MATH®

Name _____   Date _____

1. Bethany writes a set of rational numbers in increasing order. Her teacher asks her to write the absolute values of these numbers in increasing order. When her teacher checks Bethany's work, she is pleased to see that Bethany has not changed the order of her numbers. Why is this?

2. Mason was ordering the following rational numbers in math class: $-3.3, -15, -8\frac{8}{9}$.
    a. Order the numbers from least to greatest.

    b. List the order of their absolute values from least to greatest.

    c. Explain why the orderings in parts (a) and (b) are different.

EUREKA
MATH®

Lesson 12:   The Relationship Between Absolute Value and
                Order

© 2019 Great Minds®. eureka-math.org

121

1. Jessie and Makayla each have a set of five rational numbers. Although their sets are not the same, their sets of numbers have absolute values that are the same. Show an example of what Jessie and Makayla could have for numbers. Give the sets in order and the absolute values in order.

   Examples may vary. If Jessie had 2, 4, 6, 8, 10, then her order of absolute values would be the same: 2, 4, 6, 8, 10. If Makayla had the numbers −10, −8, −6, −4, −2, then her order of absolute values would also be 2, 4, 6, 8, 10.

   > Since the absolute value of a number is the distance between the number and zero on the number line, it is always a positive value. A number and its opposite have the same absolute value, so I can use any five rational numbers for Jessie's list and their opposites for Makayla's list. To put the numbers in Makayla's list in order, I remember to think of where those numbers are on the number line.

2. For each pair of rational numbers below, place each number in the Venn diagram based on how it compares to the other.

   a. −6, −1

   b. 8, −3

   > In part (a), I know −1 is greater than −6 since it's closer to 0 on the number line. I know −6 has the greater absolute value because it has a greater distance from zero. For part (b), 8 is greater than −3 and also has the larger absolute value. I can place −3 in the *None of the Above* section since it does not fit into any of the three sections of the Venn diagram.

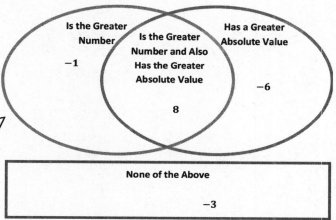

1. Micah and Joel each have a set of five rational numbers. Although their sets are not the same, their sets of numbers have absolute values that are the same. Show an example of what Micah and Joel could have for numbers. Give the sets in order and the absolute values in order.

    Enrichment Extension: Show an example where Micah and Joel both have positive and negative numbers.

2. For each pair of rational numbers below, place each number in the Venn diagram based on how it compares to the other.
    a.   −4, −8
    b.   4, 8
    c.   7, −3
    d.   −9, 2
    e.   6, 1
    f.   −5, 5
    g.   −2, 0

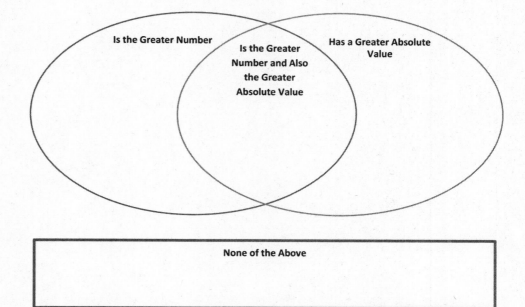

Opening Exercise

A radio disc jockey reports that the temperature outside his studio has changed 10 degrees since he came on the air this morning. Discuss with your group what listeners can conclude from this report.

Example 1: Ordering Numbers in the Real World

A $25 credit and a $25 charge appear similar, yet they are very different.

Describe what is similar about the two transactions.

How do the two transactions differ?

Exercises

1. Scientists are studying temperatures and weather patterns in the Northern Hemisphere. They recorded temperatures (in degrees Celsius) in the table below as reported in emails from various participants. Represent each reported temperature using a rational number. Order the rational numbers from least to greatest. Explain why the rational numbers that you chose appropriately represent the given temperatures.

Temperatures as Reported	8 below zero	12	−4	13 below zero	0	2 above zero	6 below zero	−5
Temperature (°C)								

2. Jami's bank account statement shows the transactions below. Represent each transaction as a rational number describing how it changes Jami's account balance. Then, order the rational numbers from greatest to least. Explain why the rational numbers that you chose appropriately reflect the given transactions.

Listed Transactions	Debit $12.20	Credit $4.08	Charge $1.50	Withdrawal $20.00	Deposit $5.50	Debit $3.95	Charge $3.00
Change to Jami's Account							

3. During the summer, Madison monitors the water level in her parents' swimming pool to make sure it is not too far above or below normal. The table below shows the numbers she recorded in July and August to represent how the water levels compare to normal. Order the rational numbers from least to greatest. Explain why the rational numbers that you chose appropriately reflect the given water levels.

Madison's Readings	$\frac{1}{2}$inch above normal	$\frac{1}{4}$inch above normal	$\frac{1}{2}$inch below normal	$\frac{1}{8}$inch above normal	$1\frac{1}{4}$inches below normal	$\frac{3}{8}$inch below normal	$\frac{3}{4}$inch below normal
Compared to Normal							

4. Changes in the weather can be predicted by changes in the barometric pressure. Over several weeks, Stephanie recorded changes in barometric pressure seen on her barometer to compare to local weather forecasts. Her observations are recorded in the table below. Use rational numbers to record the indicated changes in the pressure in the second row of the table. Order the rational numbers from least to greatest. Explain why the rational numbers that you chose appropriately represent the given pressure changes.

Barometric Pressure Change (Inches of Mercury)	Rise 0.04	Fall 0.21	Rise 0.2	Fall 0.03	Rise 0.1	Fall 0.09	Fall 0.14
Barometric Pressure Change (Inches of Mercury)							

EUREKA MATH

Example 2: Using Absolute Value to Solve Real-World Problems

The captain of a fishing vessel is standing on the deck at 23 feet above sea level. He holds a rope tied to his fishing net that is below him underwater at a depth of 38 feet.

Draw a diagram using a number line, and then use absolute value to compare the lengths of rope in and out of the water.

Example 3: Making Sense of Absolute Value and Statements of Inequality

A recent television commercial asked viewers, "Do you have over $10,000 in credit card debt?"

What types of numbers are associated with the word *debt*, and why? Write a number that represents the value from the television commercial.

Give one example of "over $10,000 in credit card debt." Then, write a rational number that represents your example.

How do the debts compare, and how do the rational numbers that describe them compare? Explain.

Lesson Summary

When comparing values in real-world situations, descriptive words help you to determine if the number represents a positive or negative number. Making this distinction is critical when solving problems in the real world. Also critical is to understand how an inequality statement about an absolute value compares to an inequality statement about the number itself.

Name _____   Date _____

1. Loni and Daryl call each other from different sides of Watertown. Their locations are shown on the number line below using miles. Use absolute value to explain who is a farther distance (in miles) from Watertown. How much closer is one than the other?

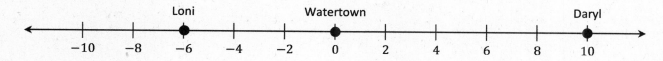

2. Claude recently read that no one has ever scuba dived more than 330 meters below sea level. Describe what this means in terms of elevation using sea level as a reference point.

Lesson 13:   Statements of Order in the Real World

1. Amy's bank account statement shows the transactions below. Write rational numbers to represent each transaction, and then order the rational numbers from greatest to least.

Listed Transactions	Debit $17.84	Credit $9.98	Charge $5.50	Withdrawal $35.00	Deposit $11.50	Debit $6.75	Charge $9.00
Change to Amy's Account	−17.84	9.98	−5.5	−35	11.5	−6.75	−9

$$11.5 > 9.98 > -5.5 > -6.75 > -9 > -17.84 > -35$$

I visualize the number line to help me determine the placement of the numbers in relation to zero.

The words "debit," "charge," and "withdrawal" all describe transactions in which money is taken out of Amy's account, decreasing its balance. I represent these transactions with negative numbers. The words "credit" and "deposit" describe transactions that will put money into Amy's account, increasing its balance, so I represent these transactions with positive numbers.

2. The fuel gauge in Holly's car says she has 29 miles to go until the tank is empty. She passed a fuel station 9 miles ago, and a sign says there is a town 15 miles ahead. If she takes a chance and drives ahead to the town and there isn't a fuel station, does she have enough fuel to go back to the fuel station? Include a diagram along a number line, and use absolute value to find your answer.

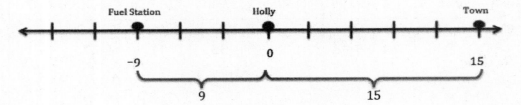

No, Holly does not have enough fuel to drive to the town and back to the gas station.

If I start at 0, where Holly is, I can think about the total number of miles from Holly to town and then how many miles it is back to the fuel station. The distance from where Holly is to town is 15 miles; then, to get to the fuel station from town, she would have to go 24 miles, which is calculated by $|15| + |-9| = 15 + 9$. The total distance is $15 + 24$, which is 39 miles. Holly would not have enough gas since she only has enough fuel for 29 miles.

She needs 15 miles worth of gas to get to town, which reduces the distance she is able to go to 14 miles ($29 - 15 = 14$). If she has to turn back and head to the fuel station, the distance is 24 miles which is calculated by $|15| + |-9| = 15 + 9$. Holly would be 10 miles short on fuel. It would be safer to go back to the fuel station without going to the town first.

     Lesson 13:   Statements of Order in the Real World   **EUREKA MATH**

© 2019 Great Minds®. eureka-math.org

1. Negative air pressure created by an air pump makes a vacuum cleaner able to collect air and dirt into a bag or other container. Below are several readings from a pressure gauge. Write rational numbers to represent each of the readings, and then order the rational numbers from least to greatest.

Gauge Readings (pounds per square inch)	25 psi pressure	13 psi vaccum	6.3 psi vaccum	7.8 psi vaccum	1.9 psi vaccum	2 psi pressure	7.8 psi pressure
Pressure Readings (pounds per square inch)							

2. The fuel gauge in Nic's car says that he has 26 miles to go until his tank is empty. He passed a fuel station 19 miles ago, and a sign says there is a town only 8 miles ahead. If he takes a chance and drives ahead to the town and there isn't a fuel station there, does he have enough fuel to go back to the last station? Include a diagram along a number line, and use absolute value to find your answer.

Example 1: The *Order* in Ordered Pairs

The first number of an ordered pair is called the _____ .

The second number of an ordered pair is called the _____ .

Example 2: Using Ordered Pairs to Name Locations

Describe how the ordered pair is being used in your scenario. Indicate what defines the first coordinate and what defines the second coordinate in your scenario.

Exercises

The first coordinates of the ordered pairs represent the numbers on the line labeled x, and the second coordinates represent the numbers on the line labeled y.

1. Name the letter from the grid below that corresponds with each ordered pair of numbers below.

    a. $(1, 4)$ b. $(0, 5)$

    c. $(4, 1)$ d. $(8.5, 8)$

    e. $(5, -2)$ f. $(5, 4.2)$

    g. $(2, -1)$ h. $(0, 9)$

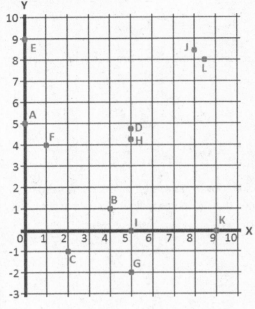

2. List the ordered pair of numbers that corresponds with each letter from the grid below.

    a.   Point M b.   Point S

    c.   Point N d.   Point T

    e.   point P f.   point U

    g.   point Q h.   point V

    i.   point R

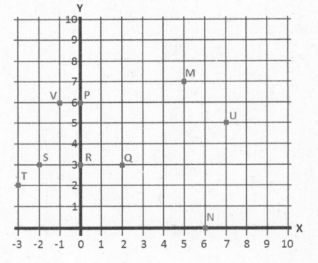

EUREKA
MATH

Lesson Summary

- The order of numbers in an ordered pair is important because the ordered pair should describe <u>one</u> location in the coordinate plane.

- The first number (called the *first coordinate*) describes a location using the horizontal direction.

- The second number (called the *second coordinate*) describes a location using the vertical direction.

Name _____   Date _____

1. On the map below, the fire department and the hospital have one matching coordinate. Determine the proper
    order of the ordered pairs in the map, and write the correct ordered pairs for the locations of the fire department
    and hospital. Indicate which of their coordinates are the same.

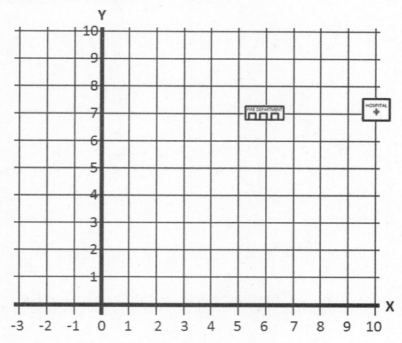

2. On the map above, locate and label the locations of each description below:

    a. The local bank has the same first coordinate as the fire department, but its second coordinate is half of the fire
        department's second coordinate. What ordered pair describes the location of the bank? Locate and label the
        bank on the map using point B.

    b. The Village Police Department has the same second coordinate as the bank, but its first coordinate is −2.
        What ordered pair describes the location of the Village Police Department? Locate and label the Village Police
        Department on the map using point P.

1. Use the set of ordered pairs below to answer each question.

$$\{(6, 15), (25, 5), (1, 2), (18, 3), (2, 17), (5, 40), (1, 7), (12, 36), (0, 9)\}$$

a. Write the ordered pair(s) whose first and second coordinate have a greatest common factor of 3.

$(6, 15)$ *and* $(18, 3)$

> I can look for ordered pairs where the first and second coordinates are multiples of 3. I can eliminate $(12, 36)$ because 12 is actually the GCF of 36, not 3.

b. Write the ordered pair(s) whose first coordinate is a factor of its second coordinate.

$(1, 2), (5, 40), (1, 7)$ *and* $(12, 36)$

> I can look for ordered pairs where the first coordinate can be multiplied by a number to get the second coordinate. So I know the first coordinate in each ordered pair is a factor of its second coordinate.
> $$1 \times 2 = 2, 5 \times 8 = 40, 1 \times 7 = 7, 12 \times 3 = 36$$

c. Write the ordered pairs(s) whose second coordinate is a prime number.

$(25, 5), (1, 2), (18, 3), (2, 17)$ *and* $(1, 7)$

> I know $5, 2, 3, 17$, and 7 are prime since they have exactly two factors, one and itself. In the other ordered pairs, the second coordinate is a composite number with three or more factors.

2. Write ordered pairs that represent the location of points A and B, where the first coordinate represents the horizontal direction, and the second coordinate represents the vertical direction.

   A: $(1, -4)$ B: $(6, 2)$

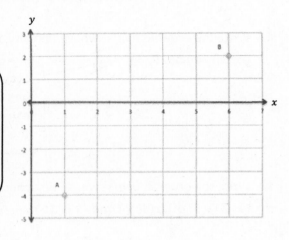

> I start at the origin $(0, 0)$. The first coordinate describes the location of the point using the horizontal direction, and the second coordinate describes the location of the point using the vertical direction. To get to point A, I can move 1 unit to the right and 4 units down. To get to point B, I can move 6 units to the right and 2 units up.

Extension:

3. Write ordered pairs of integers that satisfy the criteria in each part below. Remember that the origin is the point whose coordinates are $(0, 0)$. When possible, give ordered pairs such that (i) both coordinates are positive, (ii) both coordinates are negative, and (iii) the coordinates have opposite signs in either order.

   a. These points' vertical distance from the origin is twice their horizontal distance.

   Answers will vary; examples are $(1, 2)$, $(-3, 6)$, $(-2, -4)$.

> The x-coordinate (the 1st coordinate) represents the horizontal distance from the origin, and the y-coordinate (the 2nd coordinate) represents the vertical distance from the origin. Whatever distance I choose for the x-coordinate is half the distance of the y-coordinate since the vertical distance from the origin is twice the horizontal distance.

> Distance is always positive, so I know the point $(-3, 6)$ is 3 units from the origin when I count horizontally and 6 units from the origin when I count vertically. I have to remember to pay close attention to the signs and what they mean in the context of this problem.

   b. These points' horizontal distance from the origin is two units more than the vertical distance.

   Answers will vary; examples are $(7, 5)$, $(-7, 5)$, $(-7, -5)$, $(7, -5)$.

   c. These points' horizontal and vertical distances from the origin are equal, but only one coordinate is positive.

> For each ordered pair, the absolute value of the x-coordinate is 2 more than the absolute value of the y-coordinate.

   Answers will vary; examples are $(2, -2)$, $(-11, 11)$.

EUREKA
MATH

1. Use the set of ordered pairs below to answer each question.

    {(4, 20), (8, 4), (2, 3), (15, 3), (6, 15), (6, 30), (1, 5), (6, 18), (0, 3)}

    a. Write the ordered pair(s) whose first and second coordinate have a greatest common factor of 3.

    b. Write the ordered pair(s) whose first coordinate is a factor of its second coordinate.

    c. Write the ordered pair(s) whose second coordinate is a prime number.

2. Write ordered pairs that represent the location of points A, B, C, and D, where the first coordinate represents the horizontal direction, and the second coordinate represents the vertical direction.

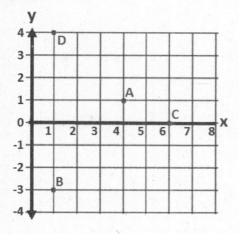

Extension:

3. Write ordered pairs of integers that satisfy the criteria in each part below. Remember that the origin is the point whose coordinates are $(0, 0)$. When possible, give ordered pairs such that (i) both coordinates are positive, (ii) both coordinates are negative, and (iii) the coordinates have opposite signs in either order.

    a. These points' vertical distance from the origin is twice their horizontal distance.

    b. These points' horizontal distance from the origin is two units more than the vertical distance.

    c. These points' horizontal and vertical distances from the origin are equal, but only one coordinate is positive.

Example 1: Extending the Axes Beyond Zero

The point below represents zero on the number line. Draw a number line to the right starting at zero. Then, follow directions as provided by the teacher.

•

Example 2: Components of the Coordinate Plane

All points on the coordinate plane are described with reference to the origin. What is the origin, and what are its coordinates?

To describe locations of points in the coordinate plane, we use _____ of numbers.

Order is important, so on the coordinate plane, we use the form (_____). The first coordinate represents the

point's location from zero on the _____- axis, and the second coordinate represents the point's location from zero on

the _____-axis.

Exercises 1–3

1. Use the coordinate plane below to answer parts (a)–(c).

    a. Graph at least five points on the x–axis, and label their coordinates.

    b. What do the coordinates of your points have in common?

    c. What must be true about any point that lies on the x-axis? Explain.

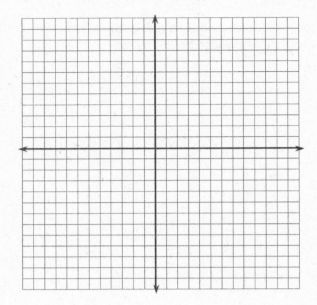

2. Use the coordinate plane to answer parts (a)–(c).

    a. Graph at least five points on the y-axis, and label their coordinates.

    b. What do the coordinates of your points have in common?

    c. What must be true about any point that lies on the y-axis? Explain.

3. If the origin is the only point with 0 for both coordinates, what must be true about the origin?

EUREKA MATH

Example 3: Quadrants of the Coordinate Plane

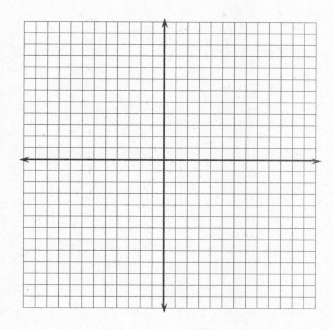

Exercises 4–6

4. Locate and label each point described by the ordered pairs below. Indicate which of the quadrants the points lie in.

   a.   (7, 2)

   b.   (3, −4)

   c.   (1, −5)

   d.   (−3, 8)

   e.   (−2, −1)

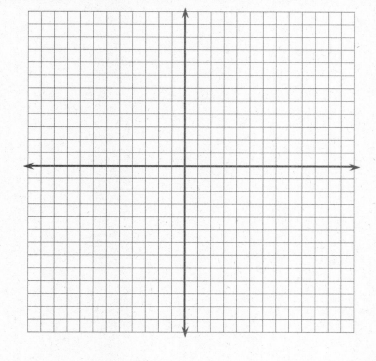

EUREKA
MATH

5. Write the coordinates of at least one other point in each of the four quadrants.

    a. Quadrant I

    b. Quadrant II

    c. Quadrant III

    d. Quadrant IV

6. Do you see any similarities in the points within each quadrant? Explain your reasoning.

EUREKA
MATH

Lesson Summary

- The x-axis and y-axis of the coordinate plane are number lines that intersect at zero on each number line.
- The axes partition the coordinate plane into four quardrants.
- Points in the coordinate plane lie either on an axis or in one of the four quadrants.

Name _____   Date _____

1. Label the second quadrant on the coordinate plane, and
    then answer the following questions:

    a. Write the coordinates of one point that lies in the
        second quadrant of the coordinate plane.

    b. What must be true about the coordinates of any
        point that lies in the second quadrant?

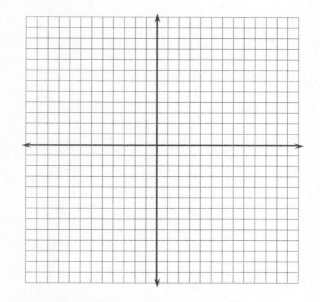

2. Label the third quadrant on the coordinate plane, and then
    answer the following questions:

    a. Write the coordinates of one point that lies in the third quadrant of the coordinate plane.

    b. What must be true about the coordinates of any point that lies in the third quadrant?

3. An ordered pair has coordinates that have the same sign. In which quadrant(s) could the point lie? Explain.

4. Another ordered pair has coordinates that are opposites. In which quadrant(s) could the point lie? Explain.

1. Name the quadrant in which each point lies. If the point does not lie in a quadrant, specify on which axis the point lies.

$(-1, 7.5)$

Quadrant II

$(7, -1)$

Quadrant IV

$(-6, -7)$

Quadrant III

$(2, 4)$

Quadrant I

$(0, 0)$

None; the point is not in a quadrant because it lies on the x-axis and y-axis

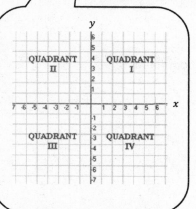

I can think about the relationship between the coordinates in each quadrant. In Quadrant I, both coordinates have positive values. In Quadrant II, the first coordinate is negative, and the second coordinate is positive. In Quadrant III, both coordinates have negative values. In Quadrant IV, the first coordinate is positive, and the second coordinate is negative.

2. Locate and label each set of points on the coordinate plane. Describe similarities of the ordered pairs in each set, and describe the points on the plane.

{(−1, 3), (−1, 5), (−1, 6), (−1, −8), (−1, −2.5)}

The ordered pairs all have x-coordinates of −1, and the points lie along a vertical line above and below (−1, 0).

I notice the x-coordinates are negative and all the same, so I know all the points will fall on a vertical line to the left of (0,0).

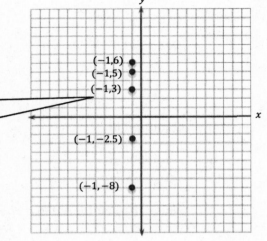

3. Locate and label at least five points on the coordinate plane that have an x-coordinate of −3.

a. What is true of the y-coordinates below the x-axis?

The y-coordinates are all negative values.

b. What is true of the y-coordinates above the x-axis?

The y-coordinates are all positive values.

c. What must be true of the y-coordinate on the x-axis?

The y-coordinates on the x-axis must be 0.

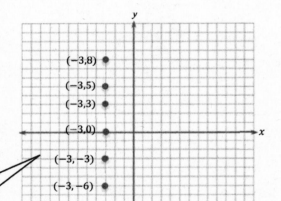

I can graph six points on the coordinate plane and look for relationships between the ordered pairs.

EUREKA
MATH

1. Name the quadrant in which each of the points lies. If the point does not lie in a quadrant, specify which axis the point lies on.

    a. $(-2, 5)$

    b. $(8, -4)$

    c. $(-1, -8)$

    d. $(9.2, 7)$

    e. $(0, -4)$

2. Jackie claims that points with the same x- and y-coordinates must lie in Quadrant I or Quadrant III. Do you agree or disagree? Explain your answer.

3. Locate and label each set of points on the coordinate plane. Describe similarities of the ordered pairs in each set, and describe the points on the plane.

    a. $\{(-2, 5) \ (-2, 2) \ (-2, 7) \ (-2, -3) \ (-2, -0.8)\}$

    b. $\{(-9, 9) \ (-4, 4) \ (-2, 2) \ (1, -1) \ (3, -3) \ (0, 0)\}$

    c. $\{(-7, -8) \ (5, -8) \ (0, -8) \ (10, -8) \ (-3, -8)\}$

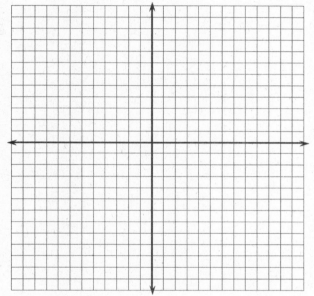

4. Locate and label at least five points on the coordinate plane that have an x-coordinate of 6.

    a. What is true of the y-coordinates below the x-axis?

    b. What is true of the y-coordinates above the x-axis?

    c. What must be true of the y-coordinates on the x-axis?

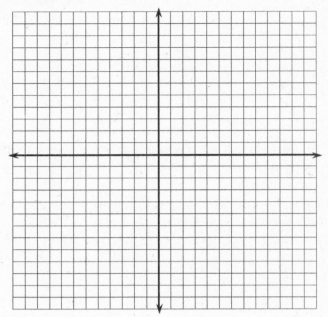

EUREKA
MATH

Opening Exercise

Give an example of two opposite numbers, and describe where the numbers lie on the number line. How are opposite numbers similar, and how are they different?

**Extending Opposite Numbers to the Coordinates of
Points on the Coordinate Plane**

Locate and lable your points on the coordinate plane to the right. For each given pair of points in the table below, record your observations and conjectures in the appropriate cell. Pay attention to the absolute values of the coordinates and where the points lie in reference to each axis.

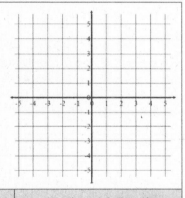

	$(3, 4)$ and $(-3, 4)$	$(3, 4)$ and $(3, -4)$	$(3, 4)$ and $(-3, -4)$
Similarities of Coordinates			
Differences of Coordinates			

Similarities in Location			
Differences in Location			
Relationship Between Coordinates and Location on the Plane			

Exercises

In each column, write the coordinates of the points that are related to the given point by the criteria listed in the first column of the table. Point $S(5, 3)$ has been reflected over the x- and y-axes for you as a guide, and its images are shown on the coordinate plane. Use the coordinate grid to help you locate each point and its corresponding coordinates.

Given Point:	$S(5, 3)$	$(-2, 4)$	$(3, -2)$	$(-1, -5)$	
The given point is reflected across the x-axis.					
The given point is reflected across the y-axis.					
The given point is reflected first across the x-axis and then across the y-axis.					
The given point is reflected first across the y-axis and then across the x-axis.					

Lesson 16:　　Symmetry in the Coordinate Plane

EUREKA MATH

1. When the coordinates of two points are (x, y) and $(-x, y)$, what line of symmetry do the points share? Explain.

2. When the coordinates of two points are (x, y) and $(x, -y)$, what line of symmetry do the points share? Explain.

Examples 2–3: Navigating the Coordinate Plane

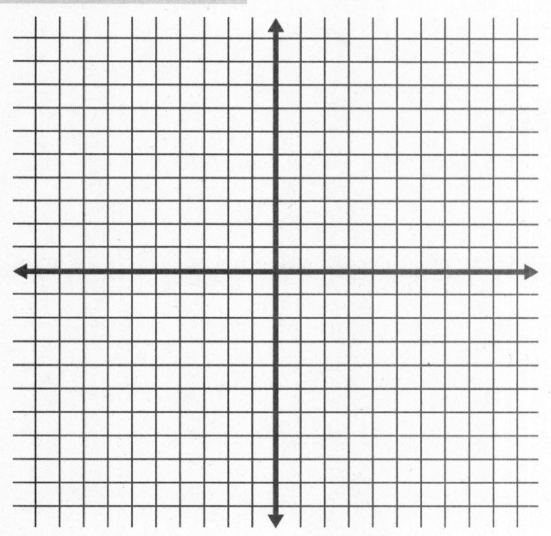

Name _____   Date _____

1. How are the ordered pairs (4, 9) and (4, −9) similar, and how are they different? Are the two points related by a reflection over an axis in the coordinate plane? If so, indicate which axis is the line of symmetry between the points. If they are not related by a reflection over an axis in the coordinate plane, explain how you know.

2. Given the point (−5, 2), write the coordinates of a point that is related by a reflection over the x- or y-axis. Specify which axis is the line of symmetry.

1. Locate a point in Quadrant *IV* of the coordinate plane. Label the point *A*, and write its ordered pair next to it.

   Answers will vary; Quadrant IV (7, −4)

   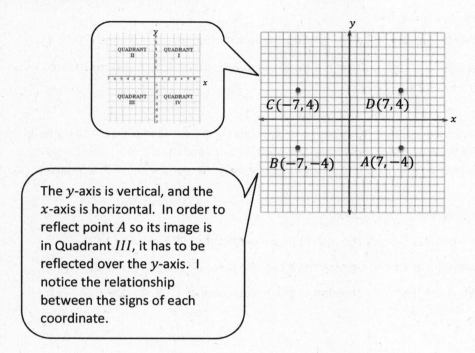

   The *y*-axis is vertical, and the *x*-axis is horizontal. In order to reflect point *A* so its image is in Quadrant *III*, it has to be reflected over the *y*-axis. I notice the relationship between the signs of each coordinate.

   a. Reflect point *A* over an axis so that its image is in Quadrant *III*. Label the image *B*, and write its ordered pair next to it. Which axis did you reflect over? What is the only difference in the ordered pairs of points *A* and *B*?

   B (−7, −4); ***reflected over the y-axis***

   The ordered pairs differ only by the sign of their x-coordintes: *A*(7, −4) ***and*** *B*(−7, −4).

b. Reflect point B over an axis so that its image is in Quadrant II. Label the image C, and write its ordered pair next to it. Which axis did you reflect over? What is the only difference in the ordered pairs of points B and C? How does the ordered pair of point C relate to the ordered pair of point A?

$C(-7, 4)$; reflected over the x-axis

The ordered pairs differ only by the signs of thier y-coordinates: $B(-7, -4)$ and $C(-7, 4)$.

The ordered pair for point C differs from the ordered pair for point A by the signs of both coordinates: $A(7, -4)$ and $C(-7, 4)$.

c. Reflect point C over an axis so that its image is in Quadrant I. Label the image D, and write its ordered pair next to it. Which axis did you reflect over? How does the ordered pair for point D compare to the ordered pair for point C? How does the ordered pair for point D compare to points A and B?

$D(7, 4)$; reflected over the y-axis again

Point D differs from point C by only the sign of its x-coordinate: $D(7, 4)$ and $C(-7, 4)$.

Point D differs from point B by the signs of both coordinates: $D(7, 4)$ and $B(-7, -4)$.

Point D differs from point A by only the sign of the y-coordinate: $D(7, 4)$ and $C(7, -4)$.

EUREKA MATH

2. Trudy listened to her teacher's directions and navigated from the point $(-3, 0)$ to $(6, -1)$. She knows that she has the correct answer, but she forgot part of the teacher's directions. Her teacher's directions included the following:

" Move 2 units up, reflect about the __?__ -axis, move down 3 units, and then move right 3 units."

Help Trudy determine the missing axis in the directions, and explain your answer.

The missing line is a reflection over the y-axis. The first move would move the location of the point to $(-3, 2)$ in Quadrant II. A reflection over the y-axis would move the location to $(3, 2)$ in Quadrant I. A move down 3 units and to the right 3 units would result in the end point $(6, -1)$.

I can visualize a coordinate plane or actually sketch one to help me with this problem. I graphed the first ordered pair at $(-3, 0)$ and then followed each direction. I can reflect the point located at 2 across the x-axis, but if I follow the rest of the steps, the result is not $(6, -1)$. I can reflect the point located at 2 across the y-axis, follow the steps, and the result is correct.

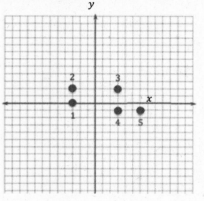

1. Locate a point in Quadrant IV of the coordinate plane. Label the point A and write its ordered pair next to it.

    a. Reflect point A over an axis so that its image is in Quadrant III. Label the image B, and write its ordered pair next to it. Which axis did you reflect over? What is the only difference in the ordered pairs of points A and B?

    b. Reflect point B over an axis so that its image is in Quadrant II. Label the image C and write its ordered pair next to it. Which axis did you reflect over? What is the only difference in the ordered pairs of points B and C? How does the ordered pair of point C relate to the ordered pair of point A?

    c. Reflect point C over an axis so that its image is in Quadrant I. Label the image D, and write its ordered pair next to it. Which axis did you reflect over? How does the ordered pair for point D compare to the ordered pair for point C? How does the ordered pair for point D compare to points A and B?

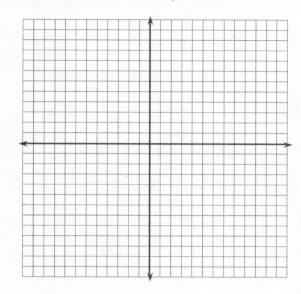

2. Bobbie listened to her teacher's directions and navigated from the point $(-1, 0)$ to $(5, -3)$. She knows that she has the correct answer, but she forgot part of the teacher's directions. Her teacher's directions included the following:

    "Move 7 units down, reflect about the ___?___ -axis, move up 4 units, and then move right 4 units."

    Help Bobbie determine the missing axis in the directions, and explain your answer.

Opening Exercise

Draw all necessary components of the coordinate plane on the blank 20 × 20 grid provided below, placing the origin at the center of the grid and letting each grid line represent 1 unit.

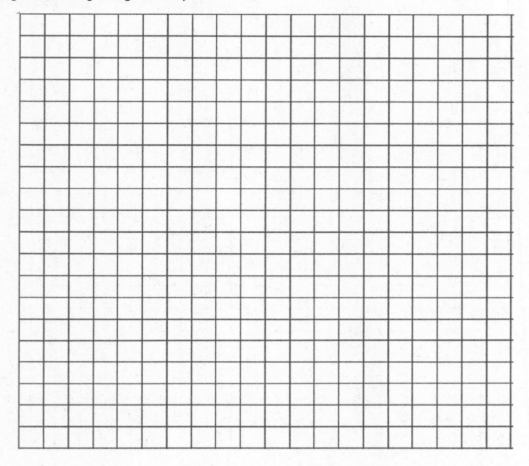

Example 1: Drawing the Coordinate Plane Using a 1 : 1 Scale

Locate and label the points {(3, 2), (8, 4), (−3, 8), (−2, −9), (0, 6), (−1, −2), (10, −2)} on the grid above.

Example 2: Drawing the Coordinate Plane Using an Increased Number Scale for One Axis

Draw a coordinate plane on the grid below, and then locate and label the following points:

$$\{(-4, 20), (-3, 35), (1, -35), (6, 10), (9, -40)\}.$$

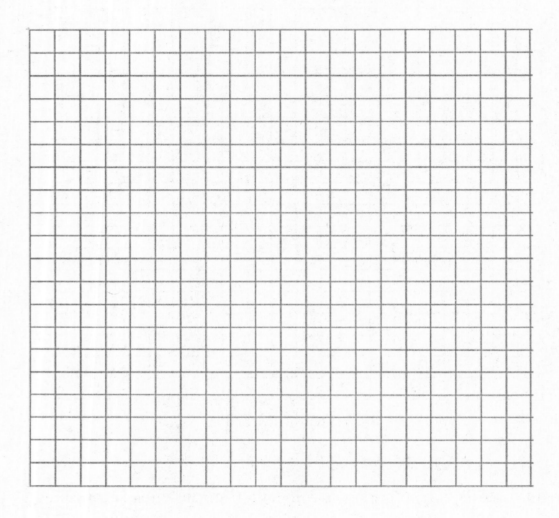

Lesson 17:   Drawing the Coordinate Plane and Points on the
Plane

**EUREKA
MATH**

Example 3: Drawing the Coordinate Plane Using a Decreased Number Scale for One Axis

Draw a coordinate plane on the grid below, and then locate and label the following points:

{(0.1, 4), (0.5, 7), (−0.7, −5), (−0.4, 3), (0.8, 1)}.

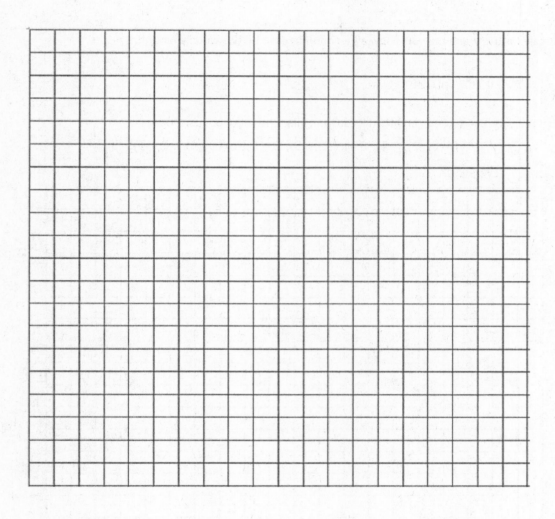

EUREKA
MATH

Lesson 17:   Drawing the Coordinate Plane and Points on the
               Plane

© 2019 Great Minds®. eureka-math.org

173

Example 4: Drawing the Coordinate Plane Using a Different Number Scale for Both Axes

Determine a scale for the x-axis that will allow all x-coordinates to be shown on your grid.

Determine a scale for the y-axis that will allow all y-coordinates to be shown on your grid.

Draw and label the coordinate plane, and then locate and label the set of points.

$$\{(-14, 2), (-4, -0.5), (6, -3.5), (14, 2.5), (0, 3.5), (-8, -4)\}$$

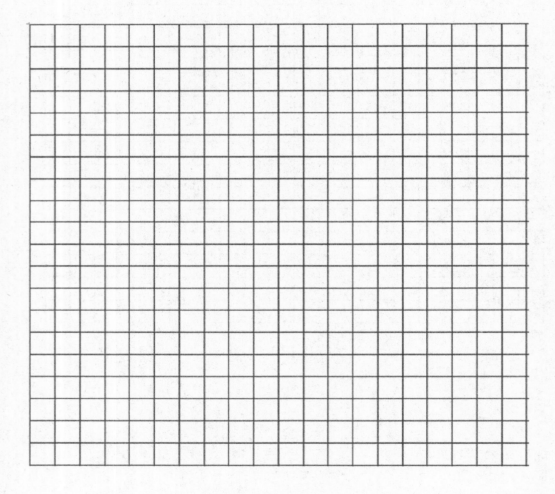

Lesson 17:   Drawing the Coordinate Plane and Points on the
              Plane

EUREKA
MATH®

Lesson Summary

- The axes of the coordinate plane must be drawn using a straightedge and labled x (horizontal axis) and y (vertical axis).

- Before assigning a scale to the axis, it is important to assess the range of values found in a set of points as well as the number of grid lines available. This allows you to determine an appropriate scale so all points can be represented on the coordinate plane that you construct.

Name _____   Date _____

Determine an appropriate scale for the set of points given below. Draw and label the coordinate plane, and then locate and label the set of points.

$$\{(10, 0.2), (-25, 0.8), (0, -0.4), (20, 1), (-5, -0.8)\}$$

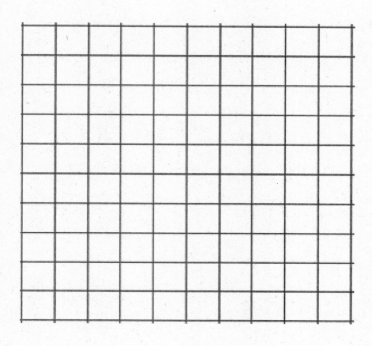

Label the coordinate plane, and then locate and label the set of points below.

$$\left\{ \begin{array}{l} (80,8),(50,5),(10,1), \\ (-30,-3),(-70,-7) \end{array} \right\}$$

To label the coordinate plane, I can look at the range of numbers for the x-coordinates and y-coordinates. In the x-coordinates, the range is from -70 to 80, so I can label the x-axis from -100 to 100. For the y-coordinates, the range is from -7 to 8, so I can label the y-axis from -10 to 10. Now, I can locate and label the points listed above.

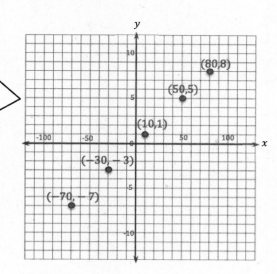

Extension:

Describe the pattern you see in the coordinates and the pattern you see in the points. Are these patterns consistent for other points, too?

The x-coordinate for each of the given points is **10** *times its y-coordinate. When I graph the points, they appear to make straight line. I check other ordered pairs with the same pattren, such as* $(-10,-1)$, $(30,3)$, *and even* $(0,0)$, *and it appears that these points are also on that line.*

1. Label the coordinate plane, and then locate and label the set of points below.

$$\begin{cases} (0.3, 0.9),\ (-0.1, 0.7),\ (-0.5, -0.1), \\ \qquad (-0.9, 0.3),\ (-0, -0.4) \end{cases}$$

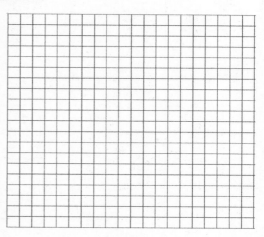

2. Label the coordinate plane, and then locate and label the set of points below.

$$\begin{cases} (90, 9),\ (-110, -11),\ (40, 4), \\ \qquad (-60, -6),\ (-80, -8) \end{cases}$$

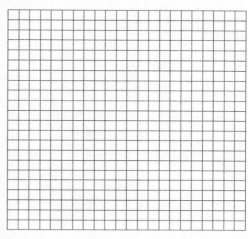

Extension:

3. Describe the pattern you see in the coordinates in Problem 2 and the pattern you see in the points. Are these patterns consistent for other points too?

Opening Exercise

Four friends are touring on motorcycles. They come to an intersection of two roads; the road they are on continues straight, and the other is perpendicular to it. The sign at the intersection shows the distances to several towns. Draw a map/diagram of the roads, and use it and the information on the sign to answer the following questions:

> Albertsville ← 8 mi.
>
> Blossville ↑ 3 mi.
>
> Cheyenne ↑ 12 mi.
>
> Dewey Falls → 6 mi.

What is the distance between Albertsville and Dewey Falls?

What is the distance between Blossville and Cheyenne?

On the coordinate plane, what represents the intersection of the two roads?

Example 1: The Distance Between Points on an Axis

Consider the points $(-4, 0)$ and $(5, 0)$.

What do the ordered pairs have in common, and what does that mean about their location in the coordinate plane?

How did we find the distance between two numbers on the number line?

Use the same method to find the distance between $(-4, 0)$ and $(5, 0)$.

Example 2: The Length of a Line Segment on an Axis

Consider the line segment with end points $(0, -6)$ and $(0, -11)$.

What do the ordered pairs of the end points have in common, and what does that mean about the line segment's location in the coordinate plane?

Find the length of the line segment described by finding the distance between its end points $(0, -6)$ and $(0, -11)$.

EUREKA
MATH

Example 3: Length of a Horizontal or Vertical Line Segment That Does Not Lie on an Axis

Consider the line segment with end points $(-3, 3)$ and $(-3, -5)$.

What do the end points, which are represented by the ordered pairs, have in common? What does that tell us about the location of the line segment on the coordinate plane?

Find the length of the line segment by finding the distance between its end points.

Exercise

Find the lengths of the line segments whose end points are given below. Explain how you determined that the line segments are horizontal or vertical.

    a. $(-3, 4)$ and $(-3, 9)$

    b. $(2, -2)$ and $(-8, -2)$

    c. $(-6, -6)$ and $(-6, 1)$

    d. $(-9, 4)$ and $(-4, 4)$

    e. $(0, -11)$ and $(0, 8)$

> **Lesson Summary**
>
> To find the distance between points that lie on the same horizontal line or on the same vertical line, we can use the same strategy that we used to find the distance between points on the number line.

EUREKA MATH

Name _____   Date _____

Determine whether each given pair of end points lies on the same horizontal or vertical line. If so, find the length of the line segment that joins the pair of points. If not, explain how you know the points are not on the same horizontal or vertical line.

   a. $(0, -2)$ and $(0, 9)$

   b. $(11, 4)$ and $(2, 11)$

   c. $(3, -8)$ and $(3, -1)$

   d. $(-4, -4)$ and $(5, -4)$

1. Find the length of the line segment with end points (6, 5) and (6, −3), and explain how you arrived at your solution.

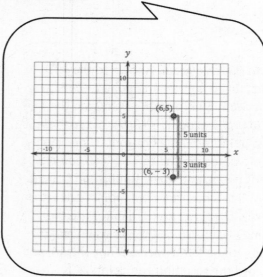

When I sketch a graph and locate the two given points, I can actually see the line segment, and it becomes easier for me to explain how to find the length.

The distance is 8 units. Both points have the same x-coordinate, so I knew they were on the same vertical line. I found the distance between the y-coordinates by counting the number of units on a vertical number line from −3 to zero and then from zero to 5, and 3 + 5 = 8.

or

I found the distance between the y-coordinates by finding the absolute value of each coordinate. |5| = 5 and |−3| = 3. The coordinates lie on opposite sides of zero, so I found the length by adding the absolute values together. Therefore, the length of a line segment with end points (6, 5) and (6, −3) is 8 units.

2. Kelly and Dave were learning partners in math class and were working independently. They each started at the point (−7, 1) and moved 6 units vertically in the plane. Each student arrived at a different end point. How is this possible? Explain and list the two different end points.

It is possible because Kelly cloud have counted up, and Dave could have counted down or vice versa. Moving 6 units in either direction vertically would generate the following possible end pints: (−7, 7) or (−7, −5).

3. The length of a line segment is 5 units. One end point of the line segment is $(-2, 9)$. Find four points that could be the other end points of the line segment.

$(-2, 14), (-2, 4), (-7, 9)$ *or* $(3, 9)$

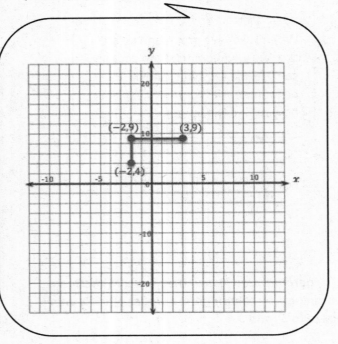

I can sketch a graph (shown to the left) and label the given end point $(-2, 9)$. I can count 5 units to the right, and the resulting end point is $(3, 9)$. Since these two points are on a horizontal line, the y-coordinates are the same. I can count 5 units down, and the resulting end point is $(-2, 4)$. Since these two points are on a vertical line, the x-coordinates are the same. I can also count 5 units up and 5 units to the left, which are not shown, and record the resulting end points.

Lesson 18:   Distance on the Coordinate Plane

EUREKA
MATH

1. Find the length of the line segment with end points (7, 2) and (−4, 2), and explain how you arrived at your solution.

2. Sarah and Jamal were learning partners in math class and were working independently. They each started at the point (−2, 5) and moved 3 units vertically in the plane. Each student arrived at a different end point. How is this possible? Explain and list the two different end points.

3. The length of a line segment is 13 units. One end point of the line segment is (−3, 7). Find four points that could be the other end points of the line segment.

Opening Exercise

In the coordinate plane, find the distance between the points using absolute value.

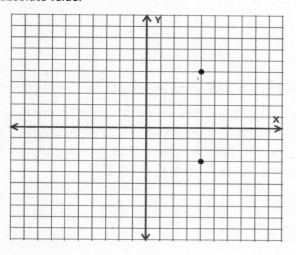

Exploratory Challenge

Exercises 1–2: The Length of a Line Segment Is the Distance Between Its End Points

1. Locate and label $(4, 5)$ and $(4, -3)$. Draw the line segment between the end points given on the coordinate plane. How long is the line segment that you drew? Explain.

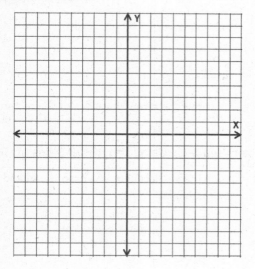

2. Draw a horizontal line segment starting at $(4, -3)$ that has a length of 9 units. What are the possible coordinates of the other end point of the line segment? (There is more than one answer.)

Which point did you choose to be the other end point of the horizontal line segment? Explain how and why you chose that point. Locate and label the point on the coordinate grid.

Exercise 3: Extending Lengths of Line Segments to Sides of Geometric Figures

3. The two line segments that you have just drawn could be seen as two sides of a rectangle. Given this, the end points of the two line segments would be three of the vertices of this rectangle.

a. Find the coordinates of the fourth vertex of the rectangle. Explain how you find the coordinates of the fourth vertex using absolute value.

b. How does the fourth vertex that you found relate to each of the consecutive vertices in either direction? Explain.

c. Draw the remaining sides of the rectangle.

Exercises 4–6: Using Lengths of Sides of Geometric Figures to Solve Problems

4. Using the vertices that you have found and the lengths of the line segments between them, find the perimeter of the rectangle.

5. Find the area of the rectangle.

Lesson 19:   Problem Solving and the Coordinate Plane

EUREKA MATH

6. Draw a diagonal line segment through the rectangle with opposite vertices for end points. What geometric figures are formed by this line segment? What are the areas of each of these figures? Explain.

Extension (If time allows): Line the edge of a piece of paper up to the diagonal in the rectangle. Mark the length of the diagonal on the edge of the paper. Align your marks horizontally or vertically on the grid, and estimate the length of the diagonal to the nearest integer. Use that estimation to now estimate the perimeter of the triangles.

Exercise 7

7. Construct a rectangle on the coordinate plane that satisfies each of the criteria listed below. Identify the coordinate of each of its vertices.

   ▪ Each of the vertices lies in a different quadrant.

   ▪ Its sides are either vertical or horizontal.

   ▪ The perimeter of the rectangle is 28 units.

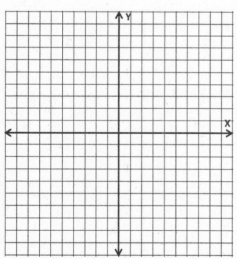

Using absolute value, show how the lengths of the sides of your rectangle provide a perimeter of 28 units.

Name _____   Date _____

1. The coordinates of one end point of a line segment are $(-2, -7)$. The line segment is 12 units long. Give three possible coordinates of the line segment's other end point.

2. Graph a rectangle with an area of 12 units² such that its vertices lie in at least two of the four quadrants in the coordinate plane. State the lengths of each of the sides, and use absolute value to show how you determined the lengths of the sides.

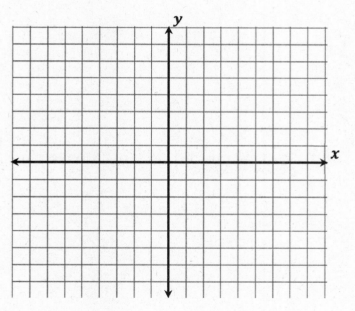

EUREKA MATH

1. One end point of a line segment is $(-2, -7)$. The length of the line segment is 4 units. Find four points that could serve as the other end point of the given line segment.

   $(-2, -11), (-2, -3), (2, -7), (-6, -7)$

   > To find four points that could be other end points for the line segment, I can move vertically or horizontally from the given point, $(-2, -7)$. If I move vertically, the y-coordinate changes, but the x-coordinate stays the same. Moving 4 units up would result in an end point of $(-2, -3)$, and moving 4 units down would result in an end point of $(-2, -11)$. If I move horizontally from the given point, the x-coordinate changes, but the y-coordinate stays the same. Moving 4 units to the right would result in an endpoint of $(2, -7)$, and moving 4 units to the left would result in an end point of $(-6, -7)$.

2. Two of the vertices of a rectangle are $(3, -4)$ and $(-5, -4)$. If the rectangle has a perimeter of 24 units, what are the coordinates of its other two vertices?

   $(-5, 0)$ and $(3, 0)$ or $(-5, -8)$ and $(3, -8)$

   > Since the two given points have the same x-coordinate, I know they are on the same horizontal line. I also know the distance from 3 to zero is 3, and the distance from zero to -5 is 5, so the total length of the line segment is 8 units. In a rectangle, there are two pairs of parallel sides, so I know the opposite side of the rectangle is also 8 units. Since the perimeter is 24 units, I can find the sum of $8 + 8$, which is 16, and subtract from 24 to determine the length of the other two sides. $24 - 16 = 8$, so the sum of the other two sides is 8. Since the remaining two sides are the same, $8 \div 2 = 4$. The length of each side is 4. From each of the given coordinates, I can count 4 units up (or down) to determine the coordinates of the other two vertices.

3. A rectangle has a perimeter of 14 units, an area of 12 square units, and sides that are either horizontal or vertical. If one vertex is the point $(-3, -2)$, and the origin is in the interior of the rectangle, find the vertex of the rectangle that is opposite $(-3, -2)$.

$(\mathbf{1}, \mathbf{1})$

I can list the factors of 12 since the area is 12 square units. The factors are $1, 2, 3, 4, 6$, and 12. In order for the perimeter of the rectangle to be 14, the sum of each half of the rectangle must be 7. So, I can look at the factors and see that $4 + 3 = 7$. Starting at the given point, I notice that a length of 3 units will not work because the origin will not be in the interior of the rectangle, so the length of the rectangle is 4 units, and the width is 3 units. I can move 4 units to the right, which will result in a vertex $(1, -2)$. Since the length is 4 units, the width is 3 units. From the point $(1, -2)$, I can move 3 units up, and the resulting vertex, which is opposite $(-3, -2)$, is $(1, 1)$.

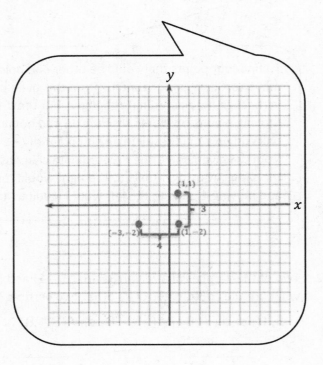

Lesson 19:   Problem Solving and the Coordinate Plane

1. One end point of a line segment is (−3, −6). The length of the line segment is 7 units. Find four points that could serve as the other end point of the given line segment.

2. Two of the vertices of a rectangle are (1, −6) and (−8, −6). If the rectangle has a perimeter of 26 units, what are the coordinates of its other two vertices?

3. A rectangle has a perimeter of 28 units, an area of 48 square units, and sides that are either horizontal or vertical. If one vertex is the point (−5, −7) and the origin is in the interior of the rectangle, find the vertex of the rectangle that is opposite (−5, −7).

Credits

Great Minds® has made every effort to obtain permission for the reprinting of all copyrighted material. If any owner of copyrighted material is not acknowledged herein, please contact Great Minds for proper acknowledgment in all future editions and reprints of this module.